BLACKST[ONE]

Blackstone's Guide to the

ROAD TRAFFIC ACT 1991

Blackstone's Guide to the
ROAD TRAFFIC ACT 1991

Simon Cooper, MA, LLB

Lecturer in Law, Newcastle Law School, University of Newcastle Upon Tyne
Consultant in Corporate Criminal Law to Pannone March Pearson, Solicitors

First published in Great Britain 1991 by Blackstone Press Limited,
9-15 Aldine Street, London W12 8AW. Telephone 081-740 1173

ISBN: 1 85431 120 4

British Library Cataloguing in Publication Data
A CIP catalogue record for this book is available from the British Library

Typeset by Style Photosetting Ltd, Mayfield, East Sussex
Printed by Ashford Colour Press, Gosport, Hampshire

Contents

Preface

Once again, road traffic law falls under the spotlight for reform. It is little more than two years since the appearance of the Road Traffic Act 1988 and the Road Traffic Offenders Act 1988. Both these Acts and others have received substantial and significant revision. Before I undertook the task of producing this text I knew that road traffic law was a complex area. As I worked my way through the provisions of this latest Act I realised that I had grossly underestimated the extent of that complexity. I have every sympathy with those whose task it is to draft legislation which adequately covers the wealth of situations which can arise out of the use of vehicular traffic. In producing this guide, I hope I will spare many of you from the traumas of undertaking tedious research through the plethora of traffic legislation. In an effort to do so, I have included a brief resume of the relevant law which surrounds the major offences revised by this Act. I hope that this will improve your understanding of the impact made by the changes this Act brings.

I must thank my friend and colleague Michael J. Allen, who undertook the task of reading the manuscript without complaint. His keen eye spared me from many an embarrassing error and, for those errors that do survive, the responsibility is mine. I would also like to express my gratitude to Heather, Alistair, Jonathan and all at Blackstone Press for their usual and ever enthusiastic encouragement.

Finally, I must mention my children, Gemma, Charlotte and Matthew. They showed great tolerance when I often absented myself in order to continue another chapter of this work.

Simon Cooper
Newcastle Law School

Chapter 1
Genesis of the Act

> *'Every day, 15 people are killed on our roads and 170 are seriously injured. It is against that background that the Government adopted the target of seeking to reduce those casualties by a full one third by the end of the decade. There is every prospect that the casualty rate can be reduced. That is what the first part of the Bill is intended to do.'* (M. Rifkind, Secretary of State for Transport, *Hansard,* Issue 1541, p. 688, 10 December 1990)

The Road Traffic Act 1991 is divided into two parts. Part I contains various amendments and additions to the existing law. Part II lays the foundations for improving traffic conditions in London. As far as practitioners are concerned, it is Part I of the Act which is the more important. It reformulates the major driving offences and revises their penalties, and creates new offences of causing death by careless driving when unfit and of endangering road users. Also introduced are two new sentencing options: rehabilitation courses for drink-drive offenders and an extended driving test for those convicted of the more serious offences. Another important feature is that, for the first time, photographic evidence will be admissible to prove offences of speeding or failing to comply with traffic signals.

Much of the Act was based on the recommendations of the Road Traffic Law Review Report compiled under the chairmanship of Dr Peter North (Dr P. North, Road Traffic Law Review Report, Dept. of Transport, (HMSO London, 1988)). The Review was set up on 31 January 1985 by the Secretaries of State for Transport and the Home Department. With the broad aim of reducing casualties on the road, the Review was asked to examine and consider what improvements might be made to certain aspects of road traffic law, including, *inter alia,* the offences of reckless and careless driving and the penalties for a number of various offences. Of the 137 recommendations made some were adopted in full, others in part, and others rejected.

The aims of the Government are admirable. With road deaths running at more than 5,000 annually, and with more than 300,000 injured, something needs to be done to reduce the unnecessary suffering that lies behind these appalling figures. The apparent apathy of the public to these dreadful statistics is something that never ceases to amaze this author. The number of people killed by individual catastrophes such as the King's Cross fire, or the Clapham train disaster, pale into insignificance when compared with the annual death toll on our roads. Yet there is no public outcry nor any significant media coverage. Whether or not the changes made by the Act will have an impact on these figures remains to be seen. This author does not believe they will. Contrary to popular belief, accidents do not 'just happen'. Accidents are caused, and for the most part are avoidable. The major causes of road accidents are irresponsible attitudes and driver behaviour. It is a peculiar characteristic of drivers that no matter what happens they see the blame as lying with someone other than themselves. An attack on a person's driving ability is somehow likened to an attack on his or her integrity as a human being. Until such attitudes can be changed and driving behaviour modified accordingly, it is likely the statistics will show little change. Making it easier to convict an offender afterwards does not prevent the accident in the first place, and until such time as the emphasis is on prevention the statistics will show little reduction.

The one measure in the Act which is likely to have the greatest effect in terms of prevention is s. 23, which provides for the admissibility of photographic evidence to prove speeding and traffic signal offences. Highly visible cameras are currently being installed at strategic sights across the country. Some will be dummies, some will not, and the cameras will be moved frequently so as to prevent drivers from learning which are in operation and which are not.

Another reason for suspecting the reduction will be minimal is this: in 1966 almost 8,000 people were killed on our roads; by 1976 this had been reduced to 6,750, and by 1986 to 5,382. During the same period the average mileage of motor vehicles had almost doubled. So in spite of the increase in vehicle mileage and the perceived problems involved in convicting bad drivers, there has been a steady reduction in the number of fatalities. There are, of course, a number of factors which must be considered when examining these statistics. There have been significant advances in medical knowledge and care, improvements in vehicle design, legislation to make the wearing of seat belts compulsory and improved road designs, all of which must have a bearing on the figures. Having said that, the number of fatalities and injuries must still be seen as unacceptably high, and anything that can be done to achieve a reduction must be welcome. This Act, however, concentrates mainly on convicting and punishing those who are actually detected offending. This may, in turn, have a deterrent effect on others and thus modify driver behaviour. However, it must surely be the likelihood of detection which carries with it the best hope of modifying driver behaviour. We only have to observe the effect that the presence of a police vehicle has on other drivers to be convinced of that. Road traffic law must be seen to provide effective sanctions against those who culpably fail to comply with it, but its lone capability to reduce fatalities and injuries must be open to some doubt.

Inextricably linked with the desire to reduce the casualty level is road safety education. After passing a test of basic competence, drivers are not required to submit themselves to any form of continued education or training. Indeed, most drivers would probably deny that they needed it. The Act gives the courts the power to order certain offenders to take extended driving tests before their driving licences are fully restored. Hitherto, the power of the courts to order a retest has been little used. The statistics have shown that this power is exercised in less than 1 per cent of cases where it is available. This stemmed from the fact that the retesting powers were not regarded as punitive. Merely being convicted of a driving offence, even if serious, was not a sufficient ground for ordering a retest. For certain offenders, retesting will now become an obligatory part of their sentence, and they will remain disqualified from driving until such time as they satisfactorily complete a retest. The test itself will either be the standard 'L' test or, for serious offenders, a new extended form of retest. The extended test will be approximately double the length of the standard 'L' test. As part of the extended test, the candidate will be required to drive for a period of time on an unrestricted dual carriageway.

Commencement Commencement of the provisions in Part I of the Act will be by Order. The main provisions of the Act are expected to commence in December 1991/January 1992.

The provisions on the retesting of drivers and courses for drink-drive offenders are expected to commence in January 1992.

The provisions relating to prohibition powers and the use of camera technology are expected to commence in April 1992.

The provisions in Part II of the Act commenced on Royal Assent, which was on 25 July 1991.

Chapter 2
New and Reformulated Offences

Sections 1, 3 and 7 amend the Road Traffic Act 1988 by creating new driving and riding offences. Section 6 creates a new offence of endangering road users. Sections 2 and 4 extend the existing offences of careless driving and driving whilst unfit through drink or drugs. Section 5 disapplies ss. 1 to 3 in respect of authorised motoring events.

2.1 Dangerous Driving

1 For sections 1 and 2 of the Road Traffic Act 1988 there shall be substituted—

> **1.** A person who causes the death of another person by driving a mechanically propelled vehicle dangerously on a road or other public place is guilty of an offence.
>
> **2.** A person who drives a mechanically propelled vehicle dangerously on a road or other public place is guilty of an offence.

Background Section 1 replaces the offences of driving recklessly or causing death by driving recklessly with new offences of driving dangerously, or causing death by driving dangerously.

The former offences involving recklessness were initially created by the Road Traffic Act 1972, and for the five-year period between 1972 and 1977 they coexisted with the offence of dangerous driving. The Criminal Law Act 1977 abolished the offence of dangerous driving, leaving prosecutors with the choice of charging offenders with either driving recklessly or the lesser offence contained in s. 3 of the 1988 Act of driving without due care and attention. (For the new definition of driving without due care and attention see 2.2 below.) This is how the situation remained for some 14 years until this Act abolished reckless driving and re-introduced the offence of dangerous driving, with a new definition being given to the word 'dangerous'. The former law was the subject of considerable criticism which was due in no small part to the definition given by the courts to the meaning of the adverb 'recklessly' (see further below, 2.1.5). Further, and perhaps because of the manner in which the courts defined recklessness, it was widely believed that many drivers who deserved to be convicted of reckless driving were, in fact, being charged with, or convicted of, only the less serious offence of careless driving. Understandably perhaps, the criticism was at its most trenchant when the driving in question had resulted in death. In such cases, the courts were constrained to impose a penalty which was in keeping with the proven fault element of carelessness, rather than a penalty which reflected the much more culpable state of recklessness. Consequently, a motorist whose driving was perceived by many to be quite dreadful, received a relatively lenient sentence, despite the fact that a death had been caused. The response has been to abolish those driving offences which involve recklessness and to replace them with new offences of driving dangerously.

2.1.1 The meaning of 'dangerous'

In previous road traffic legislation 'reckless' was not defined. This Act however, does define the meaning of the word 'dangerous', and the definition provided bears a remarkable similarity to that given by the Scottish courts when they defined the meaning of the word 'reckless'. In *Allen* v *Patterson* [1980] RTR 97, Lord Emslie, commenting on the meaning of 'recklessly', said:

> Section 2 as its language plainly, we think, suggests, requires a judgment to be made quite objectively of a particular course of driving in proved circumstances, and what the Court or a Jury has to decide, using its common sense, is whether that course of driving in these circumstances had the grave quality of recklessness. Judges and Juries will regularly understand, and Juries might well be reminded, that before they can apply the adverb 'recklessly' to the driving in question they must find that it fell far below the standard of driving expected of the competent and careful driver and that it occurred either in the face of obvious and material dangers which were or should have been observed, appreciated and guarded against, or in circumstances which showed complete disregard for any potential dangers which might result from the way in which the vehicle was being driven. It will be understood that in reaching a decision upon the critical issue a Judge or Jury will be entitled to have regard to any explanation offered by the accused driver to show that his driving in the particular circumstances did not possess the quality of recklessness at the material time.

Section 2A(1) of the 1988 Act as substituted by the 1991 Act, now provides that a person is to be regarded as driving dangerously if:

(a) the way he drives falls far below what would be expected of a competent and careful driver (the *actus reus*); and

(b) it would be obvious to a competent and careful driver that driving in that way would be dangerous (the *mens rea*).

This test requires the jury to answer two questions:

(i) Did the defendant drive his vehicle in such a way that his driving fell far below what would be expected of a competent and careful driver? This is a purely objective assessment of the defendant's standard of driving. It would be no answer for him to say, for example, that he was doing his incompetent best.
(ii) Would it be obvious to a competent and careful driver that driving in that way would be dangerous?

In each case, it is for the jury to determine what the standard of the competent and careful driver is. If the first question is answered in the negative, then the defendant is entitled to an acquittal. If, however, the jury consider that the standard of driving exhibited by the defendant did fall far below the level which would be expected of the competent and careful driver, they must then go on to consider the second question: would it have been *obvious* to the competent and careful driver that driving in that way would be dangerous?

To answer this question it is necessary to refer to the meaning of 'dangerous' as defined in s. 2A(3) of the 1988 Act. This section provides that in ss. 1 and 2 'dangerous' refers to danger either of:

(a) injury to any person, or
(b) serious damage to property.

The same section then continues that in order to determine what would be expected of, or obvious to, a competent and careful driver in a particular case, regard shall be had not only to the circumstances of which the defendant should have been aware, but also to the circumstances which are shown to have been within his knowledge. This brings an element of subjectivity into an otherwise objective test, and this element of subjectivity may work for or against the defendant. If it is demonstrated that his driving actually did fall far below the expected standard, the fact that he was aware of this and also knew his driving created a danger of injury or serious property damage, means his conviction is inevitable.

However, a situation could arise in which the defendant's driving, viewed from a purely objective stance, does fall far below the expected standard, and yet his actual knowledge might lead the jury to conclude that any competent and careful driver in those same circumstances would not perceive an obvious risk of injury or serious damage to property. Let us say that the defendant (D) is driving his car late at night on a wide and deserted stretch of road. The speed restriction is 40 mph and D is driving at 50 mph. D is approaching a set of traffic signals which is displaying red. D has a clear and unrestricted view of the junction, including the intersecting road. There is no other traffic in the vicinity of the junction, and D knows that if he continues through the red signal there is no possibility of any other road user being injured or of any property being damaged, let alone seriously damaged. Is D guilty of driving dangerously? Arguably, his conscious and deliberate decision to travel through red traffic signals at 50 mph would lead a jury to conclude that his driving was such that it fell far below the standard expected of a competent and careful driver. No competent and careful driver would ever consciously and deliberately drive at speed through a red traffic signal. But given D's actual knowledge of the circumstances existing at the time, could it be said that the competent and careful driver sharing this knowledge would regard the driving as dangerous in the sense meant by the Act? 'Dangerous' refers to a danger of injury or serious damage to property. Therefore it would be open to a jury that accepted D's version of events, to legitimately conclude that the competent and careful driver (albeit that he would not have driven in that manner in the first place) would not have perceived any obvious danger from the driving. If so, D is not guilty of driving dangerously.

Attributing the defendant's actual knowledge to the hypothetical motorist against whom he is judged mitigates the potential severity which an entirely objective test sometimes brings. As a practical consideration, if the defendant wishes to raise this issue it would seem essential that he testify and produce some credible evidence to support his contention. It may be that this will be viewed as an evidential burden which, if satisfied, would then have to be disproved by the prosecution. It would, of course, be different if the defendant, for whatever reason, did not possess knowledge that his driving was not dangerous in the relevant sense. Here, the jury would simply have to decide whether or not the competent and careful driver would have perceived an obvious danger. Similarly, if the jury decide that the competent and careful driver, even though he shares the defendant's knowledge, would nonetheless have perceived an obvious danger, the defendant is guilty. It will not avail him to protest his honest and genuine belief that no danger existed. Doubtless in the vast majority of cases the jury will not find it unduly difficult to answer both questions, but in cases where the defendant asserts an absence of any obvious danger it will be for the jury to decide whether or not any danger would have been obvious to the competent and careful motorist taking into account the defendant's actual knowledge of the circumstances.

As before, the only distinction between ss. 1 and 2 is that the former requires that a death has been caused by the dangerous driving in question. If not satisfied that the driving caused the death, the jury may return a verdict of dangerous driving (see 4.1.3 below).

2.1.2 The act of driving a defective vehicle as dangerous driving

It is a defence to the former charge of reckless driving, and also to careless driving, that the driver, without any fault, lost control of the vehicle because of a mechanical defect of which he had no knowledge and could not by the exercise of reasonable diligence discover. That doubtless remains the position under the new Act.

However, it was established under the former law that if the defendant drove a vehicle while having knowledge of a defect, he could be convicted of reckless driving (or careless driving – see below, 2.2). Reckless driving was not restricted to the actual *manner* in which the defendant drove a vehicle, but extended to cover the situation where he knowingly drove a defective vehicle. In *R* v *Crossman* [1986] RTR 49, the Court of Appeal held that a lorry driver was correctly convicted of causing death by reckless driving when his load fell off and killed a pedestrian, the driver having decided to drive even though he knew there was a risk of the load falling off and causing death or injury.

Section 2A(2) of the Road Traffic Act 1988 provides that a person is to be regarded as driving dangerously if it would be obvious to a competent and careful driver that driving the vehicle in its current state would be dangerous. In determining the state of the vehicle, s. 2A(4) enables the jury to have regard to anything attatched to or carried on or in the vehicle and the manner in which it is attatched or carried. The effect of these provisions is to preserve the law relating to the driving of vehicles which are unsafe *per se*. The jury must be satisfied that it would be obvious to a competent and careful driver that driving the vehicle in such a state would be dangerous, in the sense ascribed to that word by s. 2A(3) (see above, 2.1.1).

Similarly, in determining what would be obvious to a competent and careful driver in any particular case, regard shall be had to any circumstances of which the defendant ought to have been aware as well as to any circumstances shown to have been within his knowledge. Therefore, in deciding whether or not the danger would be obvious, the jury can vest the hypothetical competent and careful driver with the same knowledge of the circumstances that the defendant had.

There is no reference in the Act to the drunken driver being regarded as a dangerous driver simply by virtue of his drunken state. Under the former law, a driver who had taken sufficient alcohol so as to know his ability to drive was impaired, and that, accordingly, there was a risk of his causing injury, was a reckless driver if the manner of his driving created an obvious and serious risk of injury or serious damage to property (*R* v *Griffiths* [1984] Cr App R 6). The court retains a discretion to exclude evidence of intoxication if it feels the prejudicial effect outweighs the probative value. The test for exclusion of such evidence is somewhat vague. In *R* v *Thorpe* [1972] 1 WLR 342 it was stated that to be admissible, 'such evidence must tend to show that the amount of drink taken was such as would adversely affect a driver or alternatively that the driver was in fact adversely affected'.

As we have seen, the new ss. 1 and 2 of the Road Traffic Act 1988 focus on the way in which the vehicle itself is actually driven rather than the state of the driver. The only exception to this is contained in new s. 2A(2) (see above).

In future, therefore (and doubtless the practice hitherto), the drink-drive offender will have to be charged with one of the specialised drink-drive offences, which now include the new offence contained in s. 3A, of causing death by driving without due care and attention or reasonable consideration whilst being under the influence of drink or drugs. On appropriate facts, it may be possible to proceed against a drunken driver who causes death, for the common law offence of manslaughter, his intoxication being evidence of recklessness (see below, 2.1.5).

2.1.3 'Mechanically propelled vehicle' and 'public place'

As well as replacing those offences involving recklessness with offences involving dangerousness, the 1991 Act extends the new offences to any 'mechanically propelled vehicle' rather than restricting them to a 'motor vehicle'. Further, the new offences may be committed in any 'public place' as well as on 'a road'.

The reason for the change from 'motor vehicle' to 'mechanically propelled vehicle' lies in the definition of a 'motor vehicle'. The Road Traffic Act 1988, s. 185, defines a motor vehicle as 'a mechanically propelled vehicle intended or adapted for use on roads'. The offence of reckless driving could only be committed by driving a motor vehicle on a road. Therefore, a person who was driving a mechanically propelled vehicle which was not intended or adapted for use on a road was incapable of committing that offence. For example, it has been held that a 'Go-Kart' is not a motor vehicle (see *Burns* v *Currell* [1963] 2 QB 433).

The necessity to show that the vehicle was intended or adapted for use on a road has now been removed and would therefore seem to include vehicles such as 'Go-Karts'. The term 'mechanically propelled vehicle' is not specifically defined by the Act but has been held to include the following:

(a) a motor car with no engine, but with the possibility that the engine may soon be replaced (*Newberry* v *Simmonds* [1961] 2 QB 345);

(b) a vehicle which, because of mechanical failure, has broken down (*R* v *Paul* [1952] NI 61);

(c) a vehicle which is being towed and which could not be driven under its own power (*Cobb* v *Whorton* [1971] RTR 392).

Examples of other vehicles which may not have been motor vehicles but which are undoubtedly mechanically propelled vehicles would be stock cars, scrambling motor cycles, works trucks, and even tanks!

The extension of ss. 1 and 2 to include all mechanically propelled vehicles perhaps takes on a greater significance in the light of the other addition to the definition, namely, the extension of ss. 1 and 2 to public places as well as roads. The vehicles mentioned above, such as stock cars, scrambling cycles and Go-Karts, may well be found in use in public places other than roads. The drivers of these vehicles will now find themselves subject to, *inter alia*, the provisions of ss. 1 and 2. There is no requirement for these drivers to be licensed. Licences remain obligatory only for the drivers of motor vehicles on roads. However, a driver convicted of driving a mechanically propelled vehicle dangerously in a public place (albeit not a road) is liable to be disqualified and receive penalty points in the usual way.

The term 'or other public place' remains undefined. It is for the prosecution to prove that the place in question is a public place. It remains a question of degree and fact as to whether or not a place is public or private. If the public are openly invited (even upon payment), then the place is a public place. If only a restricted section of the public is invited, then it is likely that this would be deemed a private place. However, if only a restricted section of the public is excluded, it is likely to be public (see *R* v *Waters* (1963) 47 Cr App R 149). A privately owned field to which the public were invited for the purposes of watching racing was held a public place (see *R* v *Collinson* (1931) 23 Cr App R 49). The problem created by car parks will remain following this Act. If the public, in fact, have access, then the car park will be a public place even though the land in question is privately owned, for example by a brewery or publican (see *Elkins* v *Cartlidge* [1947] 1 All ER 829).

Excluded from ss. 1–3 of the 1988 Act are authorised motoring events. These events are now covered by the Road Traffic Act 1988, s. 13A, which was added by s. 5 of the 1991 Act. The section provides that a person shall not be guilty of an offence under ss. 1, 2 or 3 if he shows that he was driving in a public place other than a road in accordance with an authorisation for a motoring event given under regulations made by the Secretary of State. This section materialised at Report stage of the Bill's passage through Parliament in response to concerns expressed by various groups who take part in rallying and time trials, a sport which frequently takes place on land owned by the Forestry Commission. Such land would fall within the existing definition of 'public place', and it was felt to be inappropriate to expose competitors in such events to liability under the Act.

On a somewhat lighter note, the draftsman appears to have overlooked the purpose of the Road Traffic Act 1988, s. 189. That section deems certain mechanically propelled vehicles not to be motor vehicles. One such vehicle is what the Act describes as 'an implement for cutting grass' – in other words, a lawnmower. The Act clearly regards these as 'mechanically propelled vehicles', and now we must assume they fall within the ambit of ss. 1 and 2. The same applies to ss. 3, 3A and 4 (see 2.2 and 2.3). Of course, the usual lawnmower is controlled by a pedestrian who walks behind it. Given that the offences in ss. 1 to 4 are committed only by drivers, it would be a bold interpretation of the word 'drivers' if it were held to include those whose only sin was to race with the mower around the lawn on a Sunday afternoon! But particularly in large public areas such as parks, the mower is of the larger species, complete with seat, steering wheel and accelerator. The courts have stated that the essence of driving is the use of 'the driver's controls for the purpose of directing the movement of the vehicle' (see *R* v *MacDonagh* [1974] QB 448). Undoubtedly the driver of such a machine does just that. If the public have access to the place in question it now seems that the dangerous, careless or drunken lawnmower driver is subject to the provisions of ss. 1 to 4 of the Act.

2.1.4 Penalties for dangerous driving

Section 26 of the Act gives effect to schedule 2 which, in turn, amends the Road Traffic Offenders Act 1988, sch 2.

The maximum sentence for causing death by dangerous driving is five years' imprisonment and/or an unlimited fine. The offence is triable only on indictment.

The maximum penalty for dangerous driving is:

(a) *Summary trial:* six months' imprisonment and/or a fine subject to the statutory maximum;
(b) *On Indictment:* two years' imprisonment and/or an unlimited fine.

For the s. 1 offence (causing death by dangerous driving), in the absence of special reasons, the offender must be disqualified for a period of *not less than* two years and his licence must be endorsed with three to 11 penalty points. For the s. 2 offence (dangerous driving), in the absence of special reasons, the offender must be disqualified for a period of *not less than* 12 months and his licence must be endorsed with three to 11 penalty points. In both cases the offender must also be ordered to take an 'appropriate driving test' and will remain disqualified after the obligatory period of disqualification has been served until such test is satisfactorily completed (see below, 4.2.5.1).

2.1.5 Manslaughter

Before moving on to consider those offences which involve carelessness, it is appropriate to consider the effect of this Act as it relates to the offence of manslaughter committed by driving. Manslaughter is made out if it can be shown that the defendant either performed an unlawful and dangerous act which was likely to result in some harm, albeit not serious harm, or was reckless. Can a person be charged with manslaughter arising from the use of a mechanically propelled vehicle?

2.1.5.1 An unlawful and dangerous act If a person drives a vehicle dangerously and causes a death, thereby committing an offence under s. 1 of the 1988 Act, is he guilty of manslaughter? After all, his act is dangerous and it transgresses the law, and therefore, in a broad sense, he has behaved unlawfully. The answer stems from the case of *Andrews* v *Director of Public Prosecutions* [1937] AC 576 where Lord Atkin said:

> There is an obvious difference in the law of manslaughter between doing an unlawful act and doing a lawful act with a degree of carelessness [dangerousness] which the legislature makes criminal. If it were otherwise a man who

killed another while driving without due care and attention would *ex necessitate* commit manslaughter.

Thus the offence of manslaughter is not committed simply because the defendant has driven a vehicle in a manner which, by the standard of the competent and careful motorist, was obviously dangerous. The act must be unlawful for some reason other than that which arises out of the mere fact of driving the vehicle, namely that the defendant had the *mens rea* required for the offence which constituted the unlawful act (see further M. Allen, *Textbook on Criminal Law*, Blackstone Press, London (1991)).

2.1.5.2 Reckless manslaughter Manslaughter in this form requires that the prosecution prove that the defendant caused the death in question recklessly. The meaning of the term 'reckless' has proved to be controversial. The case of *R* v *Lawrence* [1981] 1 All ER 974, [1982] AC 510, was concerned with the definition of recklessness as used in the Road Traffic Act 1972. The leading speech was delivered by Lord Diplock, who said:

> In my view, an appropriate instruction to the jury on what is meant by driving recklessly would be that they must be satisfied of two things:
> *First*, that the defendant was in fact driving the vehicle in such a manner as to create an obvious and serious risk of causing physical injury to some other person who might happen to be using the road or of doing substantial damage to property; and
> *Second*, that in driving in that manner the defendant did so without having given any thought to the possibility of there being any such risk or, having recognised that there was some risk involved, had nonetheless gone on to take it.

The relationship between manslaughter and the then statutory offence of causing death by driving recklessly was considered in *R* v *Seymour* [1983] 2 AC 493. The House of Lords upheld a conviction for manslaughter arising from the accused's use of a lorry to shunt a car out of his path. The unfortunate driver of the car was killed. Lord Roskill said the decision in *Lawrence* applied equally to the common-law offence of manslaughter, although he added that the risk of death from the driving must be very high. The prosecution should elect which of the two charges to proceed with, and only rarely should it be necessary to charge a defendant with the offence of 'motor manslaughter'.

Although the statutory offence involving recklessness has gone, prosecutors will still be faced with the choice of preferring a charge of either causing death by driving dangerously or manslaughter. As discussed earlier, the test of dangerousness is whether the competent and careful motorist would perceive an obvious danger of injury or serious damage to property (see 2.1.1 above).

For manslaughter, the test is that set out in *Lawrence* (above), pointing out to the jury that the risk of death being caused by the manner of the driving must be very high. The fact that the offence of manslaughter can, and perhaps ought to, be charged in appropriate cases, is supported by the Road Traffic Offenders Act 1988, s. 36, as amended by the Road Traffic Act 1991, s. 32, which provides for compulsory disqualification on conviction for manslaughter for drivers of motor vehicles (see below, 4.2.3). Given that the maximum punishment for the statutory offence is five years, a charge of manslaughter will no doubt be reserved for the driver who exhibits particularly grave conduct which perhaps merits a prison sentence longer than this. A further consideration must be that if the jury acquit of manslaughter they are precluded from returning an alternative verdict of guilty to the statutory offence (see below, 4.1.3).

At 2.1.2 above, it was discussed how a person who drives a vehicle in a dangerous state can be guilty of the offence of dangerous driving. If, as a result of this type of dangerous driving a death is caused, the defendant is guilty of causing death by dangerous driving. This is in spite of the fact that ss. 1 and 2 generally focus on the actual way in which the vehicle was driven rather than on the state of the vehicle itself. But could such a driver be charged with the offence of manslaughter if, as a result of his decision to drive an unsafe vehicle, a death is caused? In *R* v *Crossman* [1986] RTR 49, the defendant was successfully prosecuted for causing death by reckless driving when he drove a vehicle with an unsafe and heavy load, knowing that there was a serious risk that it might fall off and injure someone. Given that the legal elements of manslaughter and the former s. 1 offence were the same, we can say that there was no legal reason why the defendant in *Crossman* could not have been charged with manslaughter. Similarly, there would seem to be no legal reason why today such a driver could not be charged with reckless manslaughter on appropriate facts. Obviously, the same practical considerations mentioned above would apply with equal force.

2.2 Careless and Inconsiderate Driving

2. For section 3 of the Road Traffic Act 1988 there shall be substituted—

> 3. If a person drives a mechanically propelled vehicle on a road or other public place without due care and attention, or without reasonable consideration for other persons using the road or place, he is guilty of an offence.

The North Report commented that there was a need for some type of criminal offence which caught the bad driver whose conduct fell short of that required to sustain a conviction under ss. 1 or 2. The idea that the existing s. 3 offence should be abolished or amended was mooted by the Review and three suggestions were advanced. First, restrict s. 3 to cases which involved specified consequences such as personal injury; secondly, provide a fuller definition of the phrase 'without due care' so as to make it clear that the offence is concerned with the actual quality of the driving; thirdly, create an additional offence punishable only by the fixed penalty system. In the event, none of these solutions was recommended by the Review, which itself expressed a desire for the retention of the existing offences contained in s. 3.

That is in essence what the legislature has opted to do, but with two amendments. The offences have been extended so that they apply to:

(a) 'mechanically propelled vehicles' rather than just motor vehicles (see above, 2.1.3); and
(b) 'public places' rather than just roads (see above, 2.1.3).

2.2.1 Driving a defective vehicle as careless driving

As we have seen, in relation to the offence of dangerous driving, the Act makes it clear that merely by driving a vehicle in a dangerous condition, a driver can be guilty of the offence contained in s. 2 or, if a death is caused, of the offence contained in s. 1. The new s. 3 makes no specific reference to a driver being guilty of this offence merely because the vehicle is in a poor condition. Under the former law, where no reference was made in any part of the Act to the condition of vehicles, the courts upheld convictions for careless driving where the only evidence related to the state of the vehicle rather than the quality of the driving exhibited (*Haynes* v *Swain* [1975] RTR 40). Whilst more appropriate charges may be found in the Road Vehicle (Construction and Use) Regulations 1986 or the Road Traffic Act 1988, ss. 40A and 41A (see below, 3.1), there is no reason to suppose that the courts will discontinue their interpretation that the driving of a defective vehicle is careless driving. If, however, the driver can show that the vehicle developed a mechanical defect of which he neither knew, nor could be expected to know, this will be a defence.

As stated above (at 2.1.2) the drunken driver is likely to be charged with one of the specialised drink-drive offences, but there is no reason why evidence of intoxication cannot be adduced to support a careless driving charge, subject to the court's discretion to exclude it as being of little or no probative value.

2.2.2 Penalties and alternative verdicts

The penalties for careless or inconsiderate driving remain the same. The offence may only be tried summarily, and if convicted the offender may be fined up to level 4 on the standard scale. Endorsement is obligatory with penalty points in the range of three to nine points. Disqualification is discretionary.

A defendant charged under ss. 1, 2 or 3A may, alternatively, be convicted of the offence in s. 3 (see below, 4.1.3).

2.3 Causing Death by Careless Driving When Under the Influence of Drink or Drugs

3. Before section 4 of the Road Traffic Act 1988 there shall be inserted—

3A—(1) If a person causes the death of another person by driving a mechanically propelled vehicle on a road or other public place without due care and attention, or without reasonable consideration for other persons using the road or place, and—

(a) is, at the time when he is driving, unfit to drive through drink or drugs, or

(b) he has consumed so much alcohol that the proportion of it in his breath, blood or urine at that time exceeds the prescribed limit, or

(c) he is, within 18 hours after that time, required to provide a specimen in pursuance of section 7 of this Act but without reasonable excuse fails to provide it, he is guilty of an offence.

Background This new offence was created following the recommendation of the Review. In some ways it falls short of what many people and groups would like to have seen, namely an offence of causing death by careless driving simpliciter. In particular, the Magistrates Association argued for the creation of an offence of causing death by careless driving. The main argument advanced in favour of such an offence was that it would help to meet the public sense of outrage at a death caused by someone's careless driving. Another argument was that it would fill the 'gap' between the reckless driver and the mere careless driver, a charge of simple careless driving being insufficiently serious for a driver who had caused a death. Those opposed to the creation of such an offence argued that it was wrong in principle to look at consequences rather than culpability. A crime based on consequences may well be justified where the level of culpability was also high, but not where the 'fault' element was mere carelessness. It was argued that the existence of such an offence would not deter, as most careless drivers are guilty of nothing more than minor errors of judgment which, by their nature, cannot be adequately guarded against. In the event the Review accepted the arguments advanced by the opponents and the suggestion that causing death by careless driving should be an offence was rejected.

The argument that a careless driver was not sufficiently culpable could not be advanced with such force if that driver were also intoxicated by drink or drugs at the time of the driving which caused the death. Before this Act, a driver who was intoxicated and whose careless driving resulted in a death would face a charge of driving with excess alcohol (or driving whilst unfit) together with a charge of careless driving. As we have seen, the mere presence of alcohol alone was generally insufficient to found a charge of causing death by reckless driving if the actual driving itself did not possess the qualities of recklessness. Exceptionally, and in the absence of any other explanation, recklessness could be made out where the amount of alcohol consumed must have been such that the defendant knew his ability to drive was substantially impaired and this created an obvious risk of injury to other road users, and the way he drove did in fact create such risk and cause death (*R* v *Griffiths* [1984] Cr App R 6, above 2.1.2). The difficult problem of proof remains, however, and those intoxicated drivers who make errors of judgment which amount only to carelessness would still face only charges of drink-driving and careless driving. The maximum penalty available for such an offender is six months' imprisonment, and in practice is likely to be less. The Review considered that this did not adequately reflect the gravity of the situation and concluded that the availability of a specific offence to cater for the drink-drive offender who causes death would be 'of real value', notwithstanding the arguments of linking offences to consequences.

The Review considered three possibilities before making its recommendation.

(1) Creating an offence of causing death through driving with more than the legally prescribed limit of alcohol in the body or otherwise unfit through drink or drugs. In order to secure a conviction in these circumstances, it would be necessary to show that the drinking had directly caused the death. This, it was felt, would put too onerous a burden on the prosecution who, in the absence of witnesses, would have difficulty in gathering the necessary evidence. One possibility that was mooted was to include in such an offence a rebuttable presumption that it was the driver's drinking that caused the death. A driver who had chosen to drink and then drive ought not to complain if the law required him to adduce evidence that the ensuing death was *not* caused as a result of that drinking. This was rejected on the grounds that to have such a presumption would be unusual and that it would perhaps be no easy task for a driver to adduce evidence in rebuttal.

(2) Creating an absolute offence of killing someone while driving with more than the legally prescribed limit of alcohol in the body or otherwise unfit through drink or drugs. This would absolve the prosecution from the need to prove that the intoxicating substance was a direct cause of death. A driver would be guilty merely by being unfit or over the limit in any situation in which a death resulted. The driver would not be able to defend himself by demonstrating that his driving in no sense contributed to or caused the death. This was rejected as being unduly harsh, in that a driver might well be over the prescribed limit and yet in no way cause the death in question, for example, where a pedestrian steps into the path of a vehicle and the circumstances are such that no driver could have avoided a collision irrespective of the presence of alcohol. The criminal law in general does not regard such a person as being the cause of the death (see *R* v *Dalloway* (1847) 2 Cox 273), and to adopt a contrary view in respect of road traffic offences would indeed seem unduly harsh.

(3) The final possibility that was considered, and the one that has been adopted, was to create an offence in which two legal elements have to be proved. First, that the driver was over the limit or unfit. Second, that there had been a degree of bad driving which amounted, at least, to driving without due care and attention.

2.3.1 Elements of the offence

2.3.1.1 Causing death by driving without due care and attention or without reasonable consideration for others The first element of the offence makes it necessary to prove that a death has been caused as a result of the defendant having driven either without due care and attention or without reasonable consideration for other persons using the road or place. If it cannot be shown that the defendant has driven in either of these ways, then the offence is not made out, even if he is unfit through drink or drugs or has alcohol in his body in excess of the prescribed limit. Again, this focuses on the desire to punish driving which is overtly defective rather than a driver who is simply 'unfit' but nevertheless manages (perhaps with an element of good fortune) otherwise to drive his vehicle in a proper manner. Therefore, and in accordance with the Review's wishes, a defendant will not be guilty of this offence if he drives whilst unfit but otherwise in accordance with the standard of the competent and careful motorist, even though he is involved in a collision which results in death. Section 3A assumes, perhaps rightly, that the presence of alcohol or drugs leads to the careless driving, which then is the direct cause of the death. A driver who causes death by careless driving but who is not otherwise 'unfit', will remain liable only for the simple offence of driving without due care and attention. This will expose him to an endorsement, three to nine penalty points, discretionary disqualification and a fine. The careless driver who causes death whilst unfit is liable to serve up to five years in prison, a mandatory disqualification of at least two years, three to 11 penalty points, and compulsory re-testing, regardless of whether or not the alcohol or drugs had any bearing on the careless driving (see below, 4.2.3). Even if he were to satisfy the court that the presence of the intoxicant bore no relevance to the subsequent carelessness, he will still be convicted.

The existing case law on driving without due care and attention or without reasonable consideration remains applicable for the purposes of this offence. Therefore, if the defendant (D) satisfies one of the conditions of 'unfitness', he will be guilty of this offence if death results in the following circumstances:

(a) D knows he is suseptible to blackouts and dizzy spells but nevertheless drives his vehicle and then suffers an attack (*R* v *Sibbles* [1959] Crim LR 660);
(b) D pleads his poor driving was due to inexperience commensurate with a novice driver (*McCrone* v *Riding* [1938] 1 All ER 157);
(c) D falls asleep at the wheel of his car and his car mounts the pavement (*Kay* v *Butterworth* (1946) 110 JP 75).

However, the defendant's driving ought not to be judged with the benefit of hindsight, so where a driver is confronted with a sudden emergency, the test should be whether or not it was reasonable for him to have acted as he did. Accordingly, the defendant was acquitted of careless driving where, all his lights having failed, he drove onto the hardshoulder of a motorway, colliding with an unlit stationary vehicle (*Jones* v *Chief Constable of Avon and Somerset Constabulary* [1986] RTR 259).

In many cases, the prosecution's only evidence will be that for some unknown reason the defendant's vehicle was involved in a collision or left the road etc. In the absence of any explanation from the defendant, if the conclusion must be that the defendant departed from the standard to be expected of a competent and careful driver, the court should find him guilty of careless driving (*Rabjohns* v *Burgar* [1972] Crim LR 46). Regard must be had, however, to any explanation advanced by the defendant, and, if more than fanciful, the prosecution must disprove it. It must be remembered, however, that the facts must be such that in the absence of a mechanical defect or other explanation, the only inference to be drawn is one of careless driving.

As mentioned above (see 2.2), if a person drives with knowledge of a mechanical defect, this may be sufficient to found a charge of careless driving. Therefore, if a person is 'unfit' and drives with knowledge that his vehicle has defective brakes, and as a result is unable to stop when a pedestrian emerges into his path, he will be guilty of this offence should the pedestrian die. It matters not that in all other ways, his driving was faultless.

While it may be careless driving to drive with mere knowledge of a defect, to be guilty of the offence in s. 3A it must be shown that the carelessness resulted in death. Therefore it must be shown that it was the defect in the vehicle which caused or contributed to the fatality. So if a defendant had knowledge of defective tyres and chose to drive with that knowledge, he may be guilty of careless driving. If whilst 'unfit', a driver collides with a pedestrian who has emerged suddenly into his path, he will not be guilty of this offence unless it can be shown that the defect in the tyres caused the death. The carelessness must come from the defect, namely the defective tyres, and it must be the carelessness which causes the death.

Where an information is laid alleging simply that the defendant's driving has caused inconvenience to another, it is necessary to prove that some other person has in fact been inconvenienced (*Dilkes* v *Bowman Shaw* [1981] RTR 4). For this new offence, which necessitates that a death be caused, that distinction should be academic. It is not difficult to conclude that death is an inconvenience!

The s. 3A offence extends to drivers of 'mechanically propelled vehicles' and to 'public places' other than roads (see 2.1.3 above, but see also 2.3.3 below).

2.3.1.2 Under the influence of drink or drugs The element of 'unfitness' is satisfied in any one of three situations as provided for by s. 3A(1)(a), (b) and (c):

(a) where at the time of driving a person is unfit to drive through drink or drugs;
(b) where a person has consumed so much alcohol that the proportion of it in his breath, blood or urine at the time exceeds the prescribed limit;
(c) where he is within 18 hours after the driving, required to provide a specimen in pursuance of the Act, but fails without reasonable excuse to provide it.

As mentioned above, it is not necessary for the prosecution to prove that the intoxicant caused the careless driving; it is enough that there was careless driving which caused a death and that the defendant transgressed one of the provisions contained in s. 3A(1)(a), (b) or (c). The existing law relating to driving while unfit or with excess alcohol applies. There is a wealth of case law on the drink-drive provisions, and a specialist work should be consulted for a detailed review (*Blackstone's Criminal Practice,* Part C.5, Blackstone Press (London, 1991)).

2.3.2 Failing to provide a specimen
As will be seen from s. 3A(1)(c) of the 1988 Act, a driver who, within 18 hours of the driving in question, is required to provide a specimen and, without reasonable excuse, fails to provide a specimen, is liable to be convicted as if he had provided a specimen which was over the limit. Some examples might make the effect of this provision a little clearer.

Example 1 Let us assume that D (who has been drinking) drives carelessly and causes the death of P, another road user. D panics and fails to stop at the scene. The police make efforts to trace D and do so some four hours later. They duly require him to provide a specimen and D refuses, without reasonable excuse, to do so. Applying s. 3A(1)(c), D is guilty of the offence.

Example 2 We now vary the facts above so that D is not detected for 19 hours after the offence. Again, he refuses to provide a specimen. Here, D will not be guilty of the offence in s. 3A as the request for a specimen is not made within 18 hours.

There may be some occassions where time is of the essence, as the following example illustrates:

Example 3 D kills someone by careless driving whilst 'unfit' and fails to stop at the scene. The police detect D 17 hours later. They require him to take a breath test but he refuses. As the police have reasonable cause to suspect D has alcohol in his body (they smell drink on his breath), they arrest D under the provisions of the Road Traffic Act 1988, s. 6, for failing to provide a sample of breath, or using the general power of arrest for arrestable offences contained in the Police and Criminal Evidence Act 1984, s. 24(6). Under the provisions of the Road Traffic Act 1988, s. 7(2), a requirement to provide specimens of breath can only be made at a police station. If the police make that requirement more than 18 hours after the incident in question and D refuses to provide the specimens, he will not be guilty under s. 3A(1)(c). Thus, in the example given, the police will have to get D to the police station and make the requirement within one hour. In practice, this may not be a frequently occurring situation, but it is one which nevertheless needs to be considered.

2.3.3 Vehicles covered by the offence
The offence in s. 3A(1)(a) may be committed by a person who drives a 'mechanically propelled vehicle', but s. 3A(3) provides that in relation to

s. 3A(1)(b), (c) and (2), the offence may only be committed by a person who drives a 'motor vehicle'. A 'motor vehicle' is defined as '– a mechanically propelled vehicle intended or adapted for use on roads' (Road Traffic Act 1988, s. 185, and see above, 2.1.3).

2.3.4 Penalties and alternative verdicts

The penalties for causing death by careless driving while unfit are the same as those for causing death by dangerous driving (see above, 2.1.4). If not satisfied the careless driving caused the death, the jury may return a verdict of simple careless driving. If not satisfied that there was careless driving, the jury may return an alternative verdict of either:

(a) driving while unfit through drink or drugs;, or
(b) driving with alcohol in excess of the prescribed limit; or
(c) failing to provide specimens for analysis (see below, 4.1.3).

2.4 Causing Danger to Road Users

6. Before section 23 of the Road Traffic Act 1988 there shall be inserted—

22A—(1) A person is guilty of an offence if he intentionally and without lawful authority or reasonable cause—

(a) causes anything to be on or over a road, or
(b) interferes with a motor vehicle, trailer or cycle, or
(c) interferes (directly or indirectly) with traffic equipment,

in such circumstances that it would be obvious to a reasonable person that to do so would be dangerous.

Background The Criminal Law Revision Committee in their Fourteenth Report (Cmnd. 7844 (1980)) recommended that the Offences against the Persons Act 1861, ss. 32–34, be redrafted to include endangering the safety of road users rather than merely users of railways. The draft Criminal Code prepared for the Law Commission (Law Com. No. 177 (1989)) endorsed this view in clause 86, which refers to an offence of 'endangering traffic'. The Road Traffic Review also recommended legislation be introduced to make it a criminal offence intentionally to obstruct a road or interfere with devices for the regulation of traffic where the offender is or ought to be aware that injury or damage may be caused thereby. We have probably all seen automatic traffic signals that some 'joker' has interfered with, the most common situation being that one of the signals has been altered so that an oncoming driver is confronted by both red and green lights. The instigator of this potentially lethal situation will now be caught by s. 22A. However, the section has a much broader application, and will cover a wider range of conduct beyond that already outlined.

2.4.1 Intentionally and without lawful authority or reasonable cause

The actual act which 'causes or interferes with' in the relevant way must be intentional. 'Intention' here means an intention to perform the act (i.e. the 'causing or the interfering') rather than an intention that danger should result. So if a person intentionally places a large boulder on a road, it will be no answer for him to say that he did not intend any danger of injury or serious property damage. He intentionally placed or caused the boulder to be on the road and that is that. But if he in fact places a large boulder on a road and for some reason he does not know it is a road (perhaps because it is dark and the road is unlit), it could not be said that that person intentionally placed the boulder on the road, and he would not be guilty. Even if he were reckless as to the existence of the road that would not be sufficient. The offence demands that the act which causes the 'thing' to be on or over the road is intentional. The same principle applies to the other two situations given in s. 22A(1)(b) and (c). The precise meaning of 'intention' in the criminal law has been an area of some controversy, and to delve more deeply into it here is beyond the scope of this work. For a full review of the meaning of intention a specialist work should be consulted (see M. Allen, *Textbook on Criminal Law,* Chapter 3, Blackstone Press (London, 1991)).

The offence is not made out if the defendant had lawful authority or reasonable cause to act as he did. If the presence of a lawful authority or reasonable cause is alleged, then it is submitted that the prosecution must negative its existence before a conviction can be sustained. 'Lawful authority' is a term that has presented difficulties. It does seem to suggest that there are legal exceptions to the offences created by s. 22A. Reasonable cause seems to suggest that as a matter of fact there may be occasions where the defendant ought not to be held criminally liable because, in the circumstances, it was reasonable for him to act as he did. It seems that the terms 'lawful authority' or 'reasonable cause' have 'an inbuilt elasticity which enables the courts to stretch them to cover new situations so that it is never possible to close the categories that might constitute [them]' (Smith and Hogan, *Criminal Law,* 6th edn, Butterworths (London, 1988), p. 695).

2.4.2 The three offences

The three offences created by the Road Traffic Act 1988, s. 22A, are relatively clear albeit that they are potentially very wide. Section 22A(1)(a) refers to causing *anything* to be on or over a road. Provided the requisite 'danger' is present, this could cover anything from a drawing pin to an elephant. The nature of the item is irrelevant so long as the circumstances are such that a reasonable person would perceive an obvious danger.

Section 22A(1)(b) refers to interference with a motor vehicle, trailer or cycle. There is certainly going to be some overlap between this offence and the offences of criminal damage contained in the Criminal Damage Act 1971. Having said that, not all 'damage' offences will be caught by the new endangerment offences, nor will all 'endangerment' offences be caught by the damage offences. There will, however, be occasions where the same set of facts will support a conviction for either type of offence. For example, uncoupling a brake cable undoubtedly damages a car, and a defendant would be guilty of the offence of criminal damage. Indeed, with the requisite *mens rea* the defendant would be guilty of criminal damage intending that, or being reckless as to whether or not, life was endangered. Undoubtedly he would also be guilty of the new offence contained in s. 22A(1)(b). He has interfered with a motor vehicle in such circumstances that the reasonable man would perceive an obvious danger. This overlap will enable the prosecution to choose whichever charge they believe is more appropriate to the facts.

The word 'interfere' is wider than 'damage', but while the latter includes the former the converse is not necessarily true. For example, P is driving his car along the road when D throws the contents of a bucket of water onto the windscreen causing P to lose control. The car is not damaged in any way but certainly it has been interfered with and, assuming that a reasonable person would perceive an obvious danger, the offence is complete. The same result would occur if D threw a plastic sheet onto the windscreen, temporarily obscuring P's vision. Conversely, not all offences of criminal damage to a motor vehicle will meet the requirements of the new offence. The new s. 22A offence requires 'danger', and of course D may damage P's car (say by snapping off the radio aerial) without that being in anyway dangerous to another person.

Section 22A(1)(c) refers to interfering with traffic equipment, which is further defined in s. 22A(3) to mean anything lawfully placed by a highway authority on or near a road, or a traffic sign lawfully placed on or near a road by a person other than a highway authority. This would cover such things as traffic lights (temporary or permanent), traffic cones, road signs etc. It also includes fences, barriers or lights lawfully placed on or near a road during roadworks, or so placed by a constable or other person acting under the instructions of a chief officer of police.

Any traffic equipment placed on or near a road is deemed to have been lawfully placed there unless the contrary is proved (s. 22A(4)). The Act provides that in this section 'road' does not include a footpath (s. 22A(5)). However, this will not affect the class of persons whom the section seeks to protect as the Act refers to persons on or near a road (see below, 2.4.3), and this undoubtedly includes pedestrians who use a footpath by the side of a road.

Because of the Scots law relating to malicious mischief, this offence does not apply north of the border (s. 22A(6)).

2.4.3 The meaning of dangerous

'Dangerous' is defined in the new s. 22A(2) of the Road Traffic Act 1988 in the same way as for dangerous driving (see above, 2.1.1). There must be a danger either of injury to any person or of serious damage to property. There is, however, one difference in that the danger of injury or of serious property damage is confined to persons or property while on or near a road. This is consistent with the purpose of the section which is to protect other road users.

2.4.4 Penalties
The Road Traffic Act 1991, sch. 2 amends the Road Traffic Offenders Act 1988, s. 22, by providing that the offence of causing danger to road users attracts the following maximum penalties:

(a) *Summary trial:* six months' imprisonment and/or a fine subject to the statutory maximum;
(b) *On indictment:* seven years' imprisonment and/or an unlimited fine.

2.5 Cycling Offences

Section 7 of the Act 1991 replaces the existing offence of reckless cycling with a new offence of dangerous cycling. The new cycling offence is modelled on the new offences contained in ss. 1 and 2 but with one major difference. The cycling offence (and the lesser offence of careless cycling) can only be committed on a 'road'. The existing definition of 'road' does not include footpaths, and accordingly this offence will not be committed by a cyclist who rides his machine on a footpath, no matter how dangerous his riding is. The refusal of the legislature to adopt the Review's recommendation to extend the scope of cycling offences to footpaths is, in this author's view, regrettable. Cyclists who ride on footpaths are both a nuisance and a potential danger. There can be no good reason why a cyclist who threatens or actually causes injury to others because of the dangerous manner of his riding should not attract the sanction available for dangerous cycling merely because the riding took place on a footpath rather than on a road. As before, such an offender can only be dealt with for the offence of riding on a footpath or, in appropriate cases, for an offence committed under the Offences against the Person Act 1861 or the Criminal Damage Act 1971.

2.5.1 The meaning of dangerous
'Dangerous' is defined in the new s. 28(2) of the 1988 Act in exactly the same way as for dangerous driving, (see above, 2.1.1).

2.5.2 Penalties for dangerous and careless cycling
The penalty for dangerous cycling is increased to a fine subject to level 4 on the scale.

The penalty for careless cycling is increased to a fine subject to level 3 on the scale.

A court may return an alternative verdict of careless cycling where an offender has been charged with dangerous cycling.

Chapter 3
Construction and Use, Licensing of Drivers and Information

Section 8 reforms the existing construction and use offences. Sections 9 to 16 amend the laws relating to vehicle examinations, prohibitions and unroadworthy vehicles. Sections 17 to 19 amend the laws relating to licensing, fitness to drive and disqualification. Section 20 provides an exception from the requirement for third-party insurance. Section 21 widens the circumstances in which the vehicle owner is required to provide details of the vehicle driver.

3.1 Construction and Use

Section 8(1) of the 1991 Act inserts a new s. 40A into the Road Traffic Act 1988, and in effect creates a new primary offence of using a vehicle in a dangerous condition. The offence is made out if the person *uses* or *causes* or *permits* another to use a motor vehicle or trailer on a road in one of four alternative situations (see 3.1.2).

3.1.1 Use, Cause or Permit
For convenience, a brief summary of the meanings of 'use', 'cause' and 'permit' is given here, but for a more detailed treatment, readers should refer to *Blackstone's Criminal Practice,* Part C.

3.1.1.1 Use When an offence is capable of commission by a person who uses, causes or permits, the word 'use' has a narrow meaning. The word obviously covers the actual driver of the vehicle in question, but also includes an employer of the driver when the driver is driving on the employer's business (*Windle* v *Dunning* [1968] 1 WLR 552). Where the driver is someone other than an employee, this does not amount to use by the vehicle owner even if the driver is driving at the owner's request (*Crawford* v *Haughton* [1972] 1 WLR 572).

3.1.1.2 Cause Before a person can be said to have 'caused' an offence, two elements need to be present. First, there needs to be a positive act on the part of the offender (*Price* v *Cromack* [1975] 1 WLR 988). Secondly, the offender must be shown to have had prior knowledge of the facts constituting the unlawful usage (*Ross Hillman Ltd* v *Bond* [1974] QB 435). Wilful blindness to an unlawful usage is insufficient to found a conviction for 'causing'.

3.1.1.3 Permit To 'permit' an offence also requires that prior knowledge on the part of the offender is established, but it differs from 'cause' in that there need not be a positive act on the part of the offender. A permission may be express, implied or simply acquiesed in. (A stricter interpretation has been adopted for the offence of no insurance, the *mens rea* for which is merely that the offender permitted the use of the vehicle.)

3.1.2 The four alternative situations

3.1.2.1 Condition of the vehicle The first of the four conditions provides that the offence is committed when the condition of the vehicle or trailer, or of its accessories or equipment is such that its use involves a danger of injury to any person. Commonly known as the offence of 'dangerous parts', this offence is very broad in scope. It might cover a rusted car wing which has a sharp edge, or even a loose driver's seat which might cause the vehicle driver to lose control.

3.1.2.2 Purpose for which the vehicle is used This part of the offence deals with a person who uses a vehicle for a particular purpose and the circumstances are such that its use involves a danger of injury. The offence is aimed at the person who uses a particular vehicle for a wholly unsuitable purpose and thereby causes a danger of injury. So, for example, this very broad provision would cover the driver of a motor vehicle who was towing a pedal cyclist at speed, or some similar situation. This offence will often overlap with other provisions, perhaps most frequently the provision which deals with overloading a vehicle. In such a case it will be more appropriate to charge the offence which best covers the facts.

3.1.2.3 Carrying passengers The third situation caters for the person whose use of the vehicle, because of the number of passengers carried, or because of the manner in which those passengers are carried, involves a danger of injury. The most common example would be overloading a vehicle with passengers so that the driver's control is adversely impaired. The offence would still be made out, however, if a single passenger were carried in any manner which presented the danger. For example, allowing a passenger to sit on the handlebars of a motor cycle or allowing a passenger in a car to sit on the driver's knee! Unlikely though this may seem, it is something this author has come across in practice.

3.1.2.4 The load of the vehicle The final situation covers the driver whose load in some way involves a danger of injury. If the weight, position or distribution of the vehicle's load involves such a danger, or the manner in which the load is secured involves such a danger, then the offence is made out. Perhaps the most frequent example of this offence will be where the driver of the vehicle has taken insufficient care to secure the load carried on the vehicle and, as a result, the load has fallen off.

3.1.3 Danger of injury

The common theme throughout this section is that the use of the vehicle, whether because of its condition, load, purpose or whatever, must involve a danger of injury to a person. The section does not specify that the person must be someone other than the offender but merely refers to a danger of injury to *any* person. Presumably, then, this will include the offender himself, and the offence would thus be committed where the danger of injury was only to the offender.

The meaning of 'danger' is not defined in s. 40A of the 1988 Act (unlike ss. 1 and 2). However, it would probably be safe to assume that the courts will assess any danger of injury by reference to objective standards, so that if any competent and careful motorist would have appreciated there was a danger the offence is complete. There must, however, be an actual danger of injury rather than a possible danger, although in most cases the difference will be of no consequence.

3.1.4 Requirements as to brakes, steering-gear or tyres and requirements as to weight

The Road Traffic Act 1991, s. 8(2) substitutes for the existing s. 42 of the 1988 Act a new s. 41A which creates a substantive offence relating to brakes, steering-gear and tyres. Section 41A of the 1988 Act now makes it an offence to contravene or fail to comply with a construction and use requirement as to brakes, steering-gear or tyres, or to use a vehicle or trailer on a road which does not comply with such a requirement. It is also an offence to cause or permit a vehicle or trailer to be so used.

Section 8(2) of the 1991 Act also enacts a new s. 41B, which makes it an offence to contravene or fail to comply with requirements as to any description of weight applicable to a goods vehicle or motor vehicle adapted to carry more than eight passengers. Similarly, it is an offence to use on a road, or cause or permit the use of, a vehicle which does not comply with any such requirement. Section 41B(2) re-enacts the existing defence to a charge under this new section, so essentially a driver who can prove he was travelling either to or from the nearest weighbridge is not guilty of the offence. (For a full explanation of this defence a specialist work should be consulted.)

The elevation of these offences into primary legislation reflects Parliament's belief that they are serious offences worthy of recognition in primary legislation.

3.1.5 Other construction and use requirements

The only other construction and use offence which merits attention is the newly created requirement, introduced by The Road Traffic Act 1991, sch. 4, para. 50, for the introduction of speed limiters on heavy goods vehicles. Regulations will prescribe for the fixing of speed limiters and may include provision for their inspection etc. This somewhat controversial requirement will be governed under the construction and use legislation.

Other than the amendments outlined above, the law relating to construction and use offences remains unchanged. Section 8(2) of the 1991 Act essentially preserves s. 42 of the 1988 Act, to provide that breaching construction and use provisions is an offence. For breach of the requirements (other than those falling within s. 41A or s. 41B above) the offences stem from ss. 41 and 42 and the appropriate regulations.

3.1.6 Penalties for breaching construction and use provisions

The penalties for breach of the construction and use provisions are contained in the amended Road Traffic Offenders Act 1988, sch. 2, para. 17 and are as follows:

(a) For a breach of new s. 40A or s. 41A, a fine subject to level 4 on the standard scale, unless the offence was committed in respect of a goods vehicle or a vehicle adapted to carry more than eight passengers in which case the fine is subject to level 5 on the standard scale. In all cases (subject to the exception in the Road Traffic Offenders Act 1988, s. 48) the offender's licence must be endorsed with three penalty points and he becomes liable for discretionary disqualification. Schedule 4, para. 101, of the 1991 Act inserts a new s. 48 into the Road Traffic Offenders Act 1988 which provides that, in the case of an offence committed under the Road Traffic Act 1988, s. 40A, the offender must not be disqualified nor must his licence be endorsed if he proves that he did not know, and had no reasonable cause to suspect, that the use of the vehicle involved a danger of injury to any person. Similarly, where a person is convicted of an offence under s. 41A of the 1988 Act, he must not be disqualified or have his licence endorsed if he proves that he did not know, and had no reasonable cause to suspect, that the facts were such that the offence would be committed. In both cases, the person convicted bears the burden of proof to the civil standard, i.e. on a balance of probabilities.

(b) For breach of s. 41B of the 1988 Act, a fine subject to level 5 on the standard scale.

(c) For breach of s. 42, a fine subject to level 3 on the standard scale, unless the offence was committed in respect of a goods vehicle or a vehicle adapted to carry more than eight passengers in which case the fine is subject to level 4 on the standard scale.

3.2 Vehicle Examinations and Related Offences

Sections 9 to 16 of the 1991 Act contain a number of miscellaneous provisions relating to the roadworthiness of vehicles, and prohibitions on the driving of unfit vehicles and the sale of unroadworthy vehicles. Some amendments are more important than others, for example, the extension of testing powers to cover all the construction and use requirements. In the text which follows, greater emphasis is given to the more practical and important amendments although reference will also be made to the less important amendments as appropriate.

3.2.1 Appointment of examiners and testing vehicles on roads

Section 9 repeals, *inter alia*, the Road Traffic Act 1988, s. 68(1) and (2), and provides for a new s. 66A which directs that the Secretary of State shall appoint such examiners as considered necessary for the purpose of carrying out the functions conferred on them by the Act. References to examiners in any legislation are now to be read as though they were references to examiners appointed under s. 66A.

The testing of the condition of a vehicle on a road is governed by the Road Traffic Act 1988, s. 67, and this is now amended by the 1991 Act, s. 10. The new section authorises an examiner to test a motor vehicle on a road in order to ascertain whether (a) the construction and use requirements, and (b) the requirement that the vehicle is not such that its use would involve a danger of injury, are complied with. In effect an examiner can now test on a road any vehicle to ensure compliance with the construction and use regulations and to ensure that use of the vehicle does not involve a danger of injury. The section further empowers the examiner, in pursuance of his powers under s. 67, to require the driver to comply with his reasonable instructions, and also makes clear that the examiner may drive the vehicle for the purpose of so testing it.

As before, a test may be deferred at the request of the driver of the vehicle, to be conducted at a time and in a place fixed in accordance with sch. 2. However, the right to defer a test does not apply in two situations, namely:

(a) where it appears to a constable that by reason of an accident it is desirable that the test be conducted immediately; or

(b) where a constable is of the opinion that the condition of the vehicle is so poor that it ought not to be allowed to proceed without a test being carried out.

3.2.2 Inspection of goods vehicles and public passenger vehicles

The inspection of goods vehicles and vehicles in public service is, and rightly so, thorough. Section 11 of the 1991 Act substitutes the existing s. 68 of the 1988 Act, with a new section extending the power to examine vehicles in public service and goods vehicles. Existing powers in the Public Passenger Vehicles Act 1981, s. 8(1) and (2) are abolished. The new section provides composite powers for the inspection of such vehicles. The flavour of the former s. 68 as it related to goods vehicles alone is retained, but a new s. 68(6) broadens the former provision by extending the new s. 68 to include motor vehicles which are not public service vehicles but are adapted to carry more than eight passengers (s. 68(6)(c)). However, the general power of an examiner to enter premises to inspect vehicles to which this section applies does not extend to the aforementioned vehicle in s. 68(6)(c). This undoubtedly reflects the fact that many such vehicles will be in private and not commercial ownership.

The same exemption applies to those vehicles which carry passengers under permits granted under the Transport Act 1985, ss. 19 or 22, namely vehicles used by educational bodies or in providing community bus services.

3.2.3 Prohibitions

3.2.3.1 Power to prohibit driving unfit vehicles Section 12 of the 1991 Act repeals the Public Passenger Vehicles Act 1981, s. 9, and substitutes s. 69 of the 1988 Act. Further, a new s. 69A is inserted into the 1988 Act. The broad effect of the section is to provide composite powers to prohibit the driving of unfit vehicles, and to provide powers for the removal of prohibitions. The new law extends prohibition powers to all vehicles rather than restricting them to goods vehicles and public service vehicles.

The prohibition powers will arise whenever a vehicle is inspected by a vehicle examiner under the provisions of the 1988 Act in any one of the following situations:

(a) by virtue of s. 41 (construction and use requirements);
(b) by virtue of s. 45 (tests of vehicles other than goods vehicles);
(c) by virtue of s. 49 (tests of certain classes of goods vehicles) or
(d) by virtue of ss. 61, 67, 68 or 77.

If, having inspected the vehicle under any of the above provisions, an authorised examiner is of the opinion that owing to defects in the vehicle it is, or is likely to become, unfit for service, he may prohibit the driving of that vehicle on the road absolutely. Alternatively, he may restrict it to being driven for a specific purpose only, or from being driven for a particular purpose.

Where it appears to an authorised constable that the defects would involve a danger of injury to any person, the power to prohibit is extended to that constable (s. 69(2)). An ‘authorised constable’ means a constable authorised to act for the purposes of this section by or on behalf of a chief officer of police (s. 69(9)).

A person who imposes such a prohibition must immediately give written notice to the person in charge of the vehicle at the time of the inspection, which specifies the defect(s), the nature of the prohibition, and when the prohibition comes into force (s. 69(6)). However, if the prohibition is imposed by a constable or by a vehicle examiner who considers that the defects involve a danger of injury to any person, then the prohibition takes immediate effect. This is consistent with the purpose of the section, namely to prevent ‘dangerous’ vehicles from being driven on the road. In all other cases, the prohibition shall come into force at a time specified by the examiner but, in all cases, not later than 10 days from the examination date (s. 69(4)). One point to be aware of is that a constable may be both a ‘vehicle examiner’ and an ‘authorised constable’ for the purposes of this section. In many cases he is likely to be both an examiner and authorised for the purposes of s. 69, but not necessarily so. The difference may not be academic, because a constable who is an examiner may impose a prohibition even though there is no danger of injury. A constable who is merely 'authorised' for s. 69 purposes, can impose a prohibition only where there is a danger of injury. The section goes on to provide that the constable or examiner may grant a written exemption to the notice of prohibition, allowing the vehicle to be used in such a manner, subject to any conditions and for such a purpose as the exemption may specify (s. 69(7)).

The new s. 69A empowers the examiner or constable who imposes a prohibition under s. 69, to make removal of that prohibition conditional on a further inspection. The section may be applied in four separate situations:

(a) If the vehicle prohibited is one adapted to carry more than eight passengers or is a public service vehicle, the prohibition may be imposed with a direction making it irremovable unless or until the vehicle has been inspected at an official PSV testing station.

(b) If the vehicle is one to which s. 49 of the 1988 Act applies (goods vehicles), the prohibition may be made irremovable until the vehicle has been inspected at an official testing station.

(c) If the vehicle is one to which s. 47 of the 1988 Act applies (one which requires an MOT test certificate), the prohibition may be made irremovable until the vehicle has been tested and a certificate issued.

(d) In any other case, the prohibition may be made irremovable unless and until the vehicle has been inspected in accordance with s. 72 by an examiner or authorised constable (see below, 3.2.3.3).

This is a potentially powerful sanction against the motorist who drives an unfit vehicle. To effectively deprive a motorist, even temporarily, of his vehicle (and possibly his livelihood) is a power that should be exercised only on rare occasions where the public need protection and after all other avenues have been explored.

3.2.3.2 Unfit and overloaded vehicles Section 13 of the 1991 Act amends the Road Traffic Act 1988, s. 70, by extending the power to prohibit the driving of an overloaded vehicle to any motor vehicle adapted to carry more than eight passengers. In essence, the section empowers examiners and authorised constables to prohibit the driving of such vehicles or goods vehicles that are overloaded, until the weight is reduced to a limit appropriate for the vehicle and official notification is given to the person in charge of the vehicle that it may now proceed. The examiner may give written notice to that person that the vehicle is to be removed from the road and taken to a place specified in the direction. In such a case, the prohibition shall not apply while the vehicle is being removed in accordance with the direction (s. 70(3)).

Section 14 substitutes the existing s. 71 of the 1988 Act with a new section, which makes it an offence to drive a vehicle in contravention of a prohibition imposed under ss. 69 or 70 or to cause or permit such vehicle to be driven. Further, the section makes it an offence to fail to comply within a reasonable time with a direction given under s. 70(3). The maximum penalty remains unchanged at a fine subject to level 5 on the standard scale.

3.2.3.3 Removal of prohibitions If a driver finds he has been issued with a prohibition notice, how does he go about getting that notice removed? Not easily, would seem to be the answer. Clearly, if an examiner has seen fit to impose a prohibition on a vehicle, there must be an effective means of ensuring that the driver concerned has remedied the defect which caused that notice to be issued in the first place, particularly when the defect was such that it involved a danger of injury. Section 15 of the 1991 Act substitutes a new s. 72 into the 1988 Act which deals with how, when and by whom a prohibition may be removed. Generally speaking, a prohibition which has been imposed under ss. 69 or 70 may be removed by any vehicle examiner or authorised constable if he is now satisfied that the vehicle is fit for service. However, this is made subject to the rest of s. 72 which goes on to impose certain conditions on that general power in certain circumstances:

(a) If the prohibition was imposed on a PSV or vehicle adapted to carry more than eight passengers or a goods vehicle by virtue of the provisions in s. 69A(1) or (2), the prohibition cannot be removed unless or until the vehicle has been inspected at the appropriate authorised testing station (s. 72(2)).

(b) If the prohibition was placed on a vehicle in accordance with s. 69A(3), it cannot be removed unless the vehicle undergoes the appropriate test and has been issued with a test certificate. When the vehicle has undergone a test and been issued with a test certificate, the prohibition shall be removed by such person as may be prescribed provided that any requirements relating to issue and production of the certificate have been complied with (s. 72(3)).

(c) If the prohibition was placed on a vehicle in accordance with s. 69A(4) then it shall not be removed unless or until any prescribed requirements relating to the inspection of the vehicle have been complied with (s. 72(4)).

These subsections, then, make a prohibition which was initially expressed as being irremovable precisely that. Once expressed as being irremovable, a prohibition simply cannot be lifted until the defects for which it was initially imposed have been rectified. However, s. 72(5) provides for an appeal against a refusal to remove a prohibition to the Secretary of State, who may then determine the appeal as he sees fit. The appeal must be in the prescribed manner on payment of the prescribed fee. As yet, no regulations have been made which govern the prescribed manner and fees.

3.2.4 Supplying unroadworthy vehicles

Section 16 of the 1991 Act, which broadly deals with unroadworthy vehicles amends s. 75 of the 1988 Act. A vehicle or trailer is now deemed to be in an unroadworthy condition if it is in such a condition that its use on a road would involve a danger of injury to any person (s. 75(3)(b)). A similar amendment is made to s. 75(4), which now makes it an offence to alter a motor vehicle so as to render its condition such that its use on a road would involve a danger of injury to any person. The last amendment is to the defence provided under s. 75(6). That section provides that a person shall not be convicted of an offence under the section if he proves that:

(a) it was supplied or altered for export from Great Britain; or
(b) that he had reasonable cause to believe that the vehicle would not be used on a road in Great Britain, or would not be so used until it had been put into a condition in which it might lawfully be so used.

(The further qualification in s. 75(c) has been repealed.)

The 1991 Act now inserts a new s. 75(6A) after this defence, which provides that (b) above (the reasonable cause defence) shall not apply to a person who, in the course of a trade or business, exposes a vehicle for sale, unless he also proves that he took all reasonable steps to ensure a prospective purchaser would be aware that usage in its current condition in Great Britain would be unlawful. In other words, car traders will have to prove that, as well as having had reasonable cause to believe the vehicle would not be used in Great Britain, they have also taken reasonable steps to make prospective purchasers aware that such use would be unlawful. The most obvious way for the trader to bring himself within this section would be for him to display a notice on the vehicle making it clear that using the vehicle in its current state in Great Britain would be unlawful. The new s. 75(6A) further provides that where the 'trader' offers to sell such a vehicle, he can not avail himself of the defence in (b) above unless he also proves that he took all reasonable steps to make the offeree similarly aware that using the vehicle in its present state in Great Britain would be unlawful. The remainder of s. 75 remains unaltered.

3.3 Licensing, Fitness and Effects of Disqualification

3.3.1 Licensing

The North Report recommended a number of changes in the driver licensing system to clarify certain ambiguities and to remedy inconsistencies of practice which had apparently developed. There were certainly a number of problems with the existing legislation. For example, should an under-age driver be charged with driving without a licence (Road Traffic Act 1988, s. 87) or with driving whilst disqualified (Road Traffic Act 1988, s. 103)? Is it right that a person who obtains a provisional licence but then drives in breach of the conditions may be charged with several offences, and yet the person who does not bother to get a licence at all is charged with only one offence? The 1991 Act addresses these issues and then goes on to deal with physical fitness to drive and the effects of a disqualification.

Section 17 amends s. 87(1) of the 1988 Act by removing the words 'if he is not the holder of' and substituting the words 'otherwise than in accordance with'. A similar amendment is made to s. 87(2) (allowing a person to drive without a licence) and s. 17(3) repeals s. 97(7) and 98(5) of the 1988 Act. The combined effect of these amendments is to create a single offence of driving on a road otherwise than in accordance with a licence granted for that class of vehicle. Henceforth, regardless of whether the offender has no licence at all or is a provisional licence holder failing to comply with conditions, or is driving under-age (see below 3.3.3), one offence only is committed under s. 87(1). The same is true of the 'cause or permit' offences in s. 87(2).

3.3.1.1 Penalties The offence of driving otherwise than in accordance with a licence attracts a fine subject to level 3 on the standard scale and, in a case where the offender's driving would not have been in accordance with any licence that could have been granted to him, obligatory endorsement with three to six penalty points. Where the offender's licence is obligatorily endorsed, he is also liable to discretionary disqualification. For example, if the only licence the offender could have obtained was provisional and he drove without 'L' plates as well as without a licence, he is liable to discretionary disqualification and the obligatory endorsement. This is because he is not driving in accordance with the only licence he could have obtained had he applied, that being a provisional licence.

3.3.2 Physical fitness

Section 18 of the 1991 Act makes minor amendments to the existing requirements in the 1988 Act which deal with the physical fitness of a driver. A new s. 92(10) is added to the 1988 Act and provides for an offence of driving a motor vehicle having obtained a licence by means of a false declaration as to any existing disability or prospective disability. The offence is not committed by merely obtaining the licence. It is necessary for the licence holder actually to drive a motor vehicle on a road.

Section 18(2) adds a new s. 94(3A) to the 1988 Act by providing for an offence of failing, without reasonable excuse, to notify the Secretary of State of any relevant or prospective disability during the currency of an existing licence. Again, for this offence to be complete, the licence holder must drive a motor vehicle on a road.

The final amendment to the physical fitness provisions is made by s. 18(3) which adds a new s. 94A to the 1988 Act. This now makes it an offence to drive a motor vehicle on a road if the Secretary of State has either refused to issue a licence, or revoked a licence, on grounds of physical fitness.

3.3.2.1 Penalties The penalties for offences relating to physical fitness are as follows:

(a) Driving after making a false declaration as to physical fitness; a fine subject to level 4 on the standard scale. Endorsement of the offender's licence with three to six penalty points is obligatory. Disqualification is discretionary.

(b) Driving after failing to notify the Secretary of State of a disability; a fine subject to level 3 on the standard scale. Endorsement of the offender's licence with three to six penalty points is obligatory. Disqualification is discretionary.

(c) Driving after refusal or revocation of a licence; a fine subject to level 5 on the standard scale and/or six months' imprisonment. Endorsement of the offender's licence with three to six penalty points is obligatory. Disqualification is discretionary.

3.3.3 Effects of disqualification

Section 19 of the 1991 Act substitutes a new s. 103 into the 1988 Act and creates offences of obtaining a licence or driving a motor vehicle on a road while disqualified for holding or obtaining a licence (s. 103(1)). Section 103(4) disapplies s. 103(1) in respect of a disqualification by virtue of the Road Traffic Act 1988, s. 101 (under-age drivers). In other words, the under-age driver does not commit an offence under this section, and therefore must be charged with the offence in s. 87(1) of driving otherwise than in accordance with a licence (see above, 3.3.1). As if to underline the point, the power of arrest given for the offence of driving while disqualified is also disapplied in relation to drivers disqualified by virtue of s. 101, notwithstanding that it does not exist in the first place as no offence is committed!

The power of arrest for this offence is contained in s. 103(3) of the 1988 Act, and is given to constables in uniform who have reasonable cause to suspect that a person driving a motor vehicle on a road is disqualified. A person who is disqualified for obtaining a licence by reason of s. 102 (disqualification to prevent duplication of licences), does not commit an offence of driving a motor vehicle on a road while so disqualified, but may commit the offence of obtaining a licence while so disqualified. In such circumstances, there is no power to arrest the offender (s. 103(5)).

3.3.3.1 Penalty for driving while disqualified The penalty for this offence remains unchanged, except that the number of penalty points to be endorsed on the offender's licence is now six.

3.4 Exception from Third-party Insurance

Section 20 of the 1991 Act amends s. 144 of the 1988 Act to provide that the minimum amount which must be deposited with the Accountant-General in order to gain exemption from the need for third-party insurance is increased from £15,000 to £500,000. Further provision is made to enable the Secretary of State to vary this amount by statutory instrument (s. 20(3)).

3.5 Information as to Driver Identity

Section 21 of the 1991 Act substitutes a new s. 172 into the 1988 Act and widens the circumstances in which the keeper of a vehicle is required to give details to the police of the identity of the vehicle driver. The section applies to the same offences as before but is now specifically extended to include the offence of manslaughter by the driver of a motor vehicle. As before, a person is not to be convicted of the offence in s. 172 if he shows that he did not know, and could not with reasonable diligence have ascertained, who the driver of the vehicle was (s. 172(4)).

The new section goes on to address corporate liability in the following way. By s. 172(5), where a body corporate is guilty of this offence and it is proved to have been committed with the consent or connivance of, or is attributable to neglect on the part of, a director, manager, secretary or similar officer, that officer individually is guilty of the offence as well as the body corporate and is liable to be punished accordingly.

Further, by s. 172(6), the defence in s. 172(4) is disapplied to corporate bodies (or partnerships in Scotland) unless the offender shows that no record was kept of the persons who drove the vehicle and that the failure to keep such records was reasonable. In effect, this imposes a duty on corporations to keep records of those who drive their vehicles, unless they can demonstrate that a failure to keep such records is reasonable in the circumstances.

By s. 172(11) of the 1988 Act, the same offence as that contained in s. 172(5) is enacted as it relates to Scotland (i.e., to partnerships or unincorporated associations).

3.5.1 Penalties for failing to provide information

Schedule 2 of the Road Traffic Offenders Act 1988 is amended to provide for discretionary disqualification and the mandatory endorsement of three penalty points on the offender's licence. These penalties do not apply where the offence is committed by a body corporate. Otherwise the penalties remain unchanged.

Chapter 4
Matters Arising at Trial and Penalties

Section 22 gives effect to sch. 1 of the Act. Section 23 provides for the admissibility of photographic evidence to prove speeding offences. Section 24 provides for alternative verdicts. Section 25 provides for interim disqualification. Sections 26 to 33 make miscellaneous amendments to the penalty points system and powers of courts to disqualify certain offenders. Section 34 provides for the issue of conditional fixed penalty tickets and unified procedures throughout Great Britain.

4.1 Matters Arising at Trial

4.1.1 Amendments to Road Traffic Offenders Act 1988, sch. 1

The Road Traffic Act 1991, s. 22, gives effect to sch. 1 of the Act, which makes various amendments to the Road Traffic Offenders Act 1988, sch. 1. The schedule is concerned with procedural requirements applicable in relation to certain offences. For the most part, the procedural requirements which are made the subject of amendment by this Act are concerned with the Road Traffic Offenders Act 1988, ss. 1, 6, 11 and 12(1). Section 1 of that Act requires certain offenders to be given a warning of possible prosecution, otherwise known as an NIP. Section 6 imposes a time limit within which summary proceedings for certain offences must be commenced. Section 11 provides for the admissibility of certain evidence by certificate, and s. 12(1) provides for the proof of the identity of a vehicle driver following a requirement made under Road Traffic Act 1988, s. 172 (see 3.5 above).

The Road Traffic Offenders Act 1988, sch. 1, para. 1, is now amended by the insertion of a new para. 1A which provides that the Road Traffic Offenders Act 1988, s. 1, is to be applied to offences of exceeding the speed limit. Certainly, until the passing of the 1988 Act, all offences of speeding attracted the requirement that the offender be given a notice of intended prosecution but, for some reason, this requirement was left out of the 1988 Act. This, it seems, was an oversight which has now been rectified by the 1991 Act. Now there is no doubt that speeding offenders must be given a notice in accordance with s. 1 of the 1988 Act.

Insofar as is relevant to this Act, the following new or amended offences are made subject to ss. 6, 11 or 12(1) of the Road Traffic Offenders Act 1988:

(a) *Offences subject to s. 11 only*
Causing death by careless driving when under the influence of drink or drugs.

(b) *Offences subject to ss. 11 and 12(1)*
Using vehicle in a dangerous condition etc.
Breach of requirement as to brakes, steering-gear or tyres.
Breach of requirement as to weight: goods and passenger vehicles.

(c) *Offences subject to ss. 6, 11 and 12(1)*
Driving after making a false declaration as to physical fitness.
Driving after failing to notify the Secretary of State of a disability.
Driving after refusal or revocation of a licence because of disability.

The remaining amendments are technical and not of any major significance and will not be dealt with here. They may be referred to in the appropriate schedule of the Act.

4.1.2 Photographic evidence in speeding and traffic signal offences
By s. 23 of the 1991 Act, a new s. 20 is inserted into the Road Traffic Offenders Act 1988, which provides for the admissibility of certain evidence in connection with speeding and traffic signal offences. The purpose of this section is to enable the prosecution to obtain a conviction by means of evidence produced by a camera which has been installed for the purpose of detecting speeding and other offences. The Metropolitan Police currently have 14 camera sites and have used the photographic evidence produced to support subsequent prosecutions for the offence of failing to comply with a traffic signal. Such evidence has needed to be corroborated by a witness to the offence, but this amendment will enable the prosecution to proceed solely on the basis of the photograph produced by the camera. For speeding offences, the consequences of this section will be the same; the prosecution will be able to proceed solely on the basis of the photographic evidence produced by the camera. In other cases, however, the Road Traffic Regulation Act 1984, s. 89(2), which remains unaffected, will prevent a motorist from being convicted for an offence of speeding solely on the opinion evidence of one witness.

The camera is triggered automatically when a vehicle crosses a red traffic signal, and produces a picture which records the date and time, the speed of the vehicle and the length of time that the light has been on red. Their success in contributing to a prosecution case is undeniable. The Metropolitan Police have recorded a 100 per cent conviction rate where photographic evidence has been available to the court. The first 1,000 prosecutions brought with the aid of photographic evidence have resulted in 1,000 convictions, with 980 motorists pleading guilty on the basis of the facts contained in the summons. Of the 20 who pleaded not guilty, 16 changed their pleas when shown the photograph, with the remaining four being convicted on the evidence.

Given that photographic evidence is clearly regarded as almost undeniable, it is essential that the equipment used is reliable. The new s. 20(1) of the 1988 Act provides that the record must be produced by 'a prescribed device' and must be accompanied by a certificate as to the circumstances in which the record was produced, which must be signed by a constable or other person authorised by or on behalf of the chief officer of police. Further, s. 20(4) provides that any record produced by 'a prescribed device' shall not be admissible as evidence of a fact unless the device is of a type approved by the Secretary of State. Where proceedings are brought with the aid of this section, evidence of a measurement made by a device or of the circumstances in which it was made, or that it was of a type approved, may be given by the production of a document signed in accordance with s. 20(1). Such a document which is duly signed is deemed to satisfy s. 20(1) unless the contrary be proved (s. 20(7)). In other words, where the prosecution produce a signed document stating the record was produced by a prescribed device, this is taken as conclusive evidence of that fact unless the contrary is proved by the defence. To ensure that such a document is admissible, the prosecution must serve the alleged offender with a copy at least seven days before the hearing or trial. The alleged offender may serve the prosecution with a notice (not less than three days before the hearing) requiring the attendance of the person who signed the document (s. 20(8)). Essentially this is a time-saving provision aimed at avoiding the unnecessary attendance of witnesses at court.

This section presently applies to 'speeding offences ' as defined, and to traffic light offences (s. 20(2)), but the Secretary of State may amend the list by making additions or deletions as he wishes (s. 20(3)).

4.1.3 Alternative verdicts

Section 24 of the 1991 Act substitutes the existing Road Traffic Offenders Act 1988, s. 24, with new provisions as to alternative verdicts. These are indicated as appropriate throughout the text, but for clarity and ease of reference, a full summary of the provisions is given in Table 4.1.

Table 4.1 Alternative verdicts

Offence charged under Road Traffic Act 1988	*Alternative*
Section 1 (causing death by dangerous driving)	Section 2 (dangerous driving) Section 3 (careless, and inconsiderate, driving
Section 2 (dangerous driving)	Section 3 (careless, and inconsiderate, driving)

Offence charged under Road Traffic Act 1988	*Alternative*
Section 3A (causing death by careless driving when under influence of drink or drugs)	Section 3 (careless, and inconsiderate, driving) Section 4(1) (driving when unfit to drive through drink or drugs) Section 5(1)(a) (driving with excess alcohol in breath, blood or urine) Section 7(6) (failing to provide specimen)
Section 4(1) (driving or attempting to drive when unfit to drive through drink or drugs)	Section 4(2) (being in charge of a vehicle when unfit to drive through drink or drugs)
Section 5(1)(a) (driving or attempting to drive with excess alcohol in breath, blood or urine)	Section 5(1)(b) (being in charge of a vehicle with excess alcohol in breath, blood or urine)
Section 28 (dangerous cycling)	Section 29 (careless, and inconsiderate, cycling)

A person charged under the Road Traffic Act 1988, s. 3A, cannot be convicted of any offence of attempting to drive, but a person charged with driving when unfit or with excess alcohol may be convicted of an attempt to so drive (s. 24(2) and (3)). Where a person is convicted by the Crown Court by virtue of this section, the Crown Court is limited to those powers that a magistrates' court would have had, had the offender been tried summarily (s. 24(4)). Similar provision is made for offenders in Scotland (s. 24(5)).

Where an offender is charged with the offence of manslaughter arising from his use of a motor vehicle, there is no provision for an alternative verdict of causing death by dangerous driving. One can only assume that this omission was deliberate.

4.1.4 Interim disqualification

Interim disqualifications are now dealt with in a new s. 26 of the Road Traffic Offenders Act 1988, as substituted by s. 25 of the 1991 Act. The new section makes minor amendments to the old, extending the section to cover Scotland and extending the power to disqualify to where the court has deferred sentence rather than merely where the court commits an offender for sentence.

The two main subsections of s. 26 (as they relate to England and Wales) now empower:

(a) a magistrates' court to disqualify a convicted offender when he is committed for sentence or when he is remitted to another magistrates' court under the provisions of the Magistrates' Courts Act 1980, s. 39 (remission for sentence). (In either case, the offender must have been convicted of an offence involving obligatory or discretionary disqualification, (s. 26(1));

(b) any court in England or Wales which defers sentence or which adjourns after convicting him but before dealing with him, to disqualify him until he has been dealt with for the offence: s. 26(2). Again, the offence for which he has been convicted must be one which involves obligatory or discretionary disqualification.

Similar provisions apply in Scotland (s. 26(3)).

In all cases, the maximum permitted period of disqualification under this section is six months (s. 26(4)). That is not to say that the court which disqualifies the offender under this section must specify a time limit. The court will merely disqualify until such time as the sentencing hearing takes place. Often such hearing will take place quickly, but in the unlikely event that more than six months elapses between the s. 26 order being made and that hearing, the disqualification automatically expires six months from the date of its initial imposition. In Scotland, where the offender has sentence deferred under the provisions of the Criminal Procedure (Scotland) Act 1975, ss. 219 or 432, if the period of deferral is for more than six months the disqualification under this section also takes effect for that longer period irrespective of s. 26(4) (s. 26(5)). Once a disqualification has been imposed under this section, then no further order can be made under the section (s. 26(6)).

As before, the offender must produce his licence to the court (s. 26(7)), unless the holder can satisfy the court he has applied for a new licence and it has not yet been received, or the Road Traffic Offenders Act 1988, s. 54 applies (a receipt is surrendered to the court, the licence having been taken out of the offender's possession under the fixed penalty system). Otherwise, failure to produce the licence is an offence (s. 26(8)).

At this stage the licence is not endorsed but the Secretary of State must be notified of the order, and if the sentencing court finally decides against disqualification, such notice must also be sent to the Secretary of State (s. 26(10)). If the sentencing court does disqualify, then any substantive period of disqualification imposed will be reduced by the amount of time the offender was disqualified under this section (s. 26(11)). For example, an offender is convicted of taking a motor vehicle without consent and is committed for sentence at the Crown Court. The magistrates make an order under s. 26. The defendant appears for sentence three months later and the Crown Court imposes a period of six months' disqualification. Having already served a disqualification period of three months, the defendant will in fact be eligible to drive again in a further three months' time.

4.2 Penalties

4.2.1 Miscellaneous amendments

Numerous amendments are made to the Road Traffic Offenders Act 1988, sch. 2, and these amendments are brought into force by s. 26 of the 1991 Act. Most amendments to sch. 2 are textual and self-explanatory. Those amendments to the penalties for substantive offences dealt with in this text are mentioned, as appropriate, along with that offence, but there are some changes to penalties which, because the substantive offence remains unaltered, are not mentioned in the main text. For ease of reference, the more important amendments not dealt with elsewhere are highlighted in Table 4.2, but for a complete list of the changes to penalties, readers should refer directly to sch. 2.

Table 4.2 Amended penalties

Offence	*Penalty*
Speeding offences generally	penalty points now 3 to 6, except where dealt with by fixed penalty, in which case 3
Driving without insurance	maximum penalty now increased to level 5 on the standard scale
Drive or attempt to drive when unfit	penalty points now 3 to 11
Drive or attempt to drive with excess alcohol	penalty points now 3 to 11
Failing to provide specimen for analysis	penalty points now 3 to 11
Motor racing on public ways	penalty points now 3 to 11
Carrying passenger on motorcycle contrary to s. 23	penalty points now 3
Driving with uncorrected defective eyesight	penalty points now 3
Failing to stop or report after accident	penalty points now 5 to 10, maximum sentence now 6 months, or fine subject to level 5 on the standard scale, or both.

4.2.2 Penalty points to be attributed and counted

4.2.2.1 Points to be attributed The existing Road Traffic Offenders Act 1988, s. 28, is replaced by a new section which is created by s. 27 of the 1991 Act. Whilst largely re-enacting its predecessor, there is one important change. This affects s. 28(4) which provides that where a person is convicted of two or more offences committed on the same occasion and which involve obligatory endorsement, the total number of points to be attributed to them is the highest number that would be attributed on a conviction for one of them. In other words, only that offence which attracts the highest number of points will be attributed, no matter how many offences qualify for endorsement.

Example 1 If D is dealt with for speeding and is also found to be driving without insurance, then these are offences which are committed on the same occasion. The speeding offence carries up to six penalty points and the insurance offence up to eight points. If the court considers that five points would be appropriate for the speeding offence and that eight points would be appropriate for the insurance offence, then only those eight points (being the higher of the two) will be endorsed on the licence. No points will be attributed for the speeding offence.

Example 2 If D is dealt with for speeding and is also found to be driving otherwise than in accordance with a licence, then these offences are committed on the same occasion. If the court specifies five points for the speeding offence and considers that only four points should be awarded for the licensing offence, then only the five points for the speeding offence will be endorsed on the licence.

The new s. 28(5) provides that where s. 28(4) above applies, the court may, if it thinks fit, determine that that subsection shall not apply. So, in our examples above, the court may, if it so deems, order that D's licence be endorsed with 13 and nine points respectively. In Example 1, this will render D liable for disqualification under the Road Traffic Offenders Act 1988, s. 35 and the 'totting-up' system. Whenever a court decides that s. 28(4) shall not apply, it is obliged to state its reasons in open court, and such reasons must be entered in the register of proceedings (s. 28(6)).

As before, a person who is convicted of aiding, abetting, counselling or procuring, or inciting commission of an offence involving obligatory disqualification, is liable to have 10 points endorsed on his licence (s. 28(2)). Where an offender is dealt with under the fixed penalty system, his licence will be endorsed with the specified number of points in sch. 2 for the offence as dealt with by that system. Therefore, a person dealt with for speeding under the fixed penalty system will have three points endorsed on his licence and will not be exposed to the risk of having six points endorsed on his licence. If the offence carries variable points and sch. 2 does not specify the number of points to be attributed under the fixed penalty system, then an offender dealt with in this way will have the lowest number of points in the variable range endorsed on his licence (s. 28(3)).

4.2.2.2 Points to be counted Section 28 of the 1991 Act substitutes a new s. 29 into the Road Traffic Offenders Act 1988. The new section provides that where a person is convicted of an offence or offences which oblige the court to endorse his licence, the following penalty points will be taken into account:

(a) all those points which are attributed to the offence or offences in respect of which he now stands convicted, but disregarding the points for any offence in respect of which an order under s. 34 is made;

(b) all those points that were previously endorsed on his licence, unless the offender has, since that previous occasion and before this conviction, been disqualified under s. 35 (the totting-up provisions).

The existing law is preserved in that if any of the offences was committed more than three years before another, the penalty points in respect of that offence shall not be added to those in respect of the other (s. 29(2)).

The new s. 29 makes an important change to the former law relating to disqualifications. Broadly speaking, offenders who are disqualified from driving for specific offences, for example drink-driving, will no longer have the penalty points from previous unrelated offences removed from their licence.

An example may serve to illustrate the effect of the new s. 29. It will be assumed that the offences were all committed within a three-year period and therefore s. 29(2) does not apply.

Example D has eight penalty points on his licence. On 1 February 1991, he is convicted of drink-driving. The court disqualifies him for 12 months by virtue of the Road Traffic Offenders Act 1988, s. 34. Particulars of the disqualification are endorsed on his licence. (There is no power to endorse the licence with penalty points if at the same time a disqualification is imposed.)

On 1 March 1992 D is now convicted of speeding. The court imposes four penalty points. In deciding whether or not D is liable for disqualification, the court will take account of the initial eight points, and the four points endorsed on this occasion. Therefore, D has 12 penalty points and is now liable for disqualification under s. 35.

Before these amendments, the defendant would have had his licence 'wiped clean' of penalty points when he was disqualified for the drink-drive offence on 1 February 1991. Now, the only time that previous penalty points are 'removed' from the equation is where they were awarded in respect of an offence committed more than three years before another, or where the offender is disqualified under the provisions of the Road Traffic Offenders Act 1988, s. 35. In the example above, if the court disqualifies the defendant under s. 35 (as they must in the absence of mitigating circumstances), this will then have the effect of wiping the licence clean of penalty points. A disqualification under the provisions of s. 34 no longer has this effect.

4.2.3 Disqualification for certain offences

The Road Traffic Offenders Act 1988, s. 34, has been mentioned above as it relates to the penalty points system (4.2.2.2). Section 29 of the 1991 Act makes amendments to s. 34 in respect of minimum periods of disqualification for certain offences. The main changes are the inclusion of the new offence under the Road Traffic Act 1988, s. 3A and increase of the minimum period of disqualification for the new offence of causing death by dangerous driving (formerly reckless driving). Again, reference is also made to these provisions in that part of the text which discusses the substantive offence, but they are included together in Table 4.3 for clarity and ease of reference. (All new offences are included here, together with the more common offences involving obligatory disqualification).

Table 4.3 Disqualification periods

	Offence	*Obligatory disqualification period*
(a)	Manslaughter arising from the use of a motor vehicle	2 years
(b)	Causing death by dangerous driving	2 years
(c)	Causing death by careless driving while unfit	2 years
(d)	Dangerous driving	12 months
(e)	Drive or attempt to drive while unfit	12 months
(f)	Drive or attempt to drive with excess alcohol	12 months
(g)	Failing to provide a specimen after driving or attempting to drive	12 months

In cases (c), (e), (f) or (g) above, the minimum period is increased to three years if the offender has been convicted of any such offence within the 10 years preceding commission of the offence for which he now stands convicted (s. 34(3)). Where an offender has been convicted of an offence involving obligatory disqualification, and it is not in catergory (c), (e), (f) or (g) above, and he has previously been disqualified on more than one occasion for a fixed period of 56 days or more within the three years preceding the offence, he must be disqualified for a minimum period of two years (s. 34(4)). This latter provision, however, does not apply in respect of disqualifications imposed under s. 26 (interim disqualifications) or under the Power of Criminal Courts Act, s. 44, (disqualification on conviction on indictment of an offence carrying at least two years' imprisonment if committed with a motor vehicle), or under the Theft Act 1968, ss. 12 or 25, or attempts thereat. The same exemption is conferred in respect of equivalent offences committed in Scotland.

Note also that offenders convicted of causing death by dangerous driving or dangerous driving *must* be disqualified under s. 36 until they have passed an appropriate driving test (see below, 4.2.6).

Finally, the power to impose a discretionary disqualification is given by s. 34(2) of the Road Traffic Offenders Act 1988, which provides that where a person is convicted of an offence involving discretionary disqualification, and either (a) the penalty points to be taken into account number fewer than 12, or (b) the offence does not involve obligatory endorsement, the court may disqualify him for such period as it thinks fit.

4.2.4 Courses for drink-drive offenders

A new s. 34A is inserted into the Road Traffic Offenders Act 1988 by virtue of s. 30 of the 1991 Act, and provides, in general, for drink-drive offenders to attend rehabilitative courses. Attendance on such a course is not considered a punitive part of the sentencing element for the particular offence; an offender cannot be compelled to attend. The aim of the course is driver education and rehabilitation. The benefit for those offenders who attend such a course will be to have their period of disqualification reduced, enabling them to return to the road sooner than those who do not attend such a course. The effectiveness of the scheme is to be tested for an experimental period in selected areas and, presumably, if successful will be extended nationwide.

This is the first time that legislation has provided a specific statutory power to enable a court to order an offender, with his consent, to undertake a course of training. However, the idea is not entirely novel. Some courts have required offenders to take courses as part of the terms of a probation order. One such course was provided for offenders who committed traffic offences while impaired by alcohol. The course was aimed at the drink-drive offender who had not previously been convicted of a similar offence, and was designed to change the offender's attitude towards drinking and subsequent behaviour. The course comprised discussions, projects, role playing exercises, and was provided by the Hampshire Probation Service in association with the University of Southampton. This course drew upon the experience of a similar course run in Germany.

Many foreign countries place a strong emphasis on driver retraining/retesting, and the subsequent reduction in reconvictions provides strong evidence that such schemes have a considerable impact on offenders. Monitored experiments in the Federal Republic of Germany (as it then was) and in California revealed, in comparison with a control group:

(a) a 60 per cent reduction in reconvictions within two years for first time drink-drive offenders;
(b) a 25 per cent reduction in reconvictions within three years for the multiple drink-drive offenders; and
(c) a 10 per cent reduction in reconvictions within one year for other non-drink offenders.

The Road Traffic Law Review considered the merits of a retraining scheme and recommended an experimental period during which certain offenders could be ordered to undergo retraining. It was further recommended that such a scheme should be based on the Hampshire Probation Service Course and restricted to selected courts within the experimental area until or unless a national scheme is implemented. The Executive response has been s. 34A.

4.2.4.1 Reduced period of disqualification for attending a course The Road Traffic Offenders Act 1988, s. 34A(1) provides that its provisions will apply where an offender has been convicted of:

(a) causing death by careless driving when under the influence of drink or drugs;

(b) driving or being in charge when under the influence of drink or drugs;
(c) driving or being in charge with excess alcohol; or
(d) failing to provide a specimen under the Road Traffic Act 1988, s. 7.

In addition, the court must make an order under Road Traffic Offenders Act 1988, s. 34, disqualifying the offender for a period of not less than 12 months.

If this section does apply, then the court may make an order which reduces the period of disqualification if, by a specified date, the offender satisfactorily completes an approved course (s. 34A(2)). The length of the reduction in the period of disqualification will differ according to how long a period was imposed on the offender in the first place. However, s. 34A(3) provides that the minimum period of reduction will be three months and not more than one-quarter of the unreduced period. In effect, this means a 25 per cent reduction in the length of disqualification. Most commonly, a drink-drive offender is disqualified for 12 months and will therefore be eligible for a three month reduction. The offender disqualified for two years will be eligible for a six-month reduction.

4.2.4.2 Conditions for attending a course Simply being convicted of a relevant offence and disqualified for at least 12 months does not mean that an offender will then automatically qualify for attendance on a course and a reduction in the period of disqualification. The Road Traffic Offenders Act 1988 imposes certain conditions and restrictions on the ability of a court to make a s. 34A order.

These conditions are contained in s. 34A(4), and the first is that the court must be satisfied that a place on the course will be available for the offender. The Hampshire Course, mentioned in 4.2.4, catered for eight offenders on each occasion it was run. Each course comprised attendance at eight weekly sessions, followed by a 10-week individual programme. Thus the course took at least 18 weeks to complete. At present, it is not known how many places will be available on the proposed courses, nor is their duration or content certain, but it may well become the case that some offenders will lose out on the benefits of attendance simply because there are no places available when their cases come before the court. Secondly, the court cannot make an order unless the offender is at least 17 years old. Thirdly, the court must explain in plain language, the effect of the order, the fees for attendance, and that those fees must be paid before the course begins. As yet, the fees for the course are unknown. The final condition is that the offender must agree to the order being made. This is, perhaps, the most important condition of the four. Whether or not an offender choses to attend the course (and the choice is clearly his) will really depend on how important it is to him that his licence is returned at the earliest available opportunity. Is three months really going to make that much difference to him? Another factor will surely be whether or not the driver concerned recognises or appreciates that he might benefit from retraining. A common fault of many drivers is that they believe they are perfect, never succumbing to error, and that whatever happens it is the other driver's fault and not theirs. I have encountered an endless number of motorists who, even in the face of overwhelming evidence to the contrary, simply refuse to concede any suggestion that they have been guilty of poor driving. Even following conviction for sometimes serious offences, I have encountered offenders who still insist it was the other driver's fault and not theirs. This peculiar attitude of the British motorist may well lead to resistance to any suggested course of retraining. At the end of the experimental period (scheduled to be 1997) it might need to be considered whether or not attendance on a course of retraining should be compulsory if, in the court's opinion, the offender requires retraining. A greater benefit for the convicted drink-drive offender who attends a course, might be available in the form of reduced insurance premiums upon restoration of his licence. If insurers consider this possibility, a reduced premium might prove a bigger incentive to attend a course than any reduction in the period of disqualification.

The only other general condition relating to the order is that the latest date for completion of a course must be *at least* two months before the last day of the reduced period of disqualification (e.g. for a 12-month disqualification which is reduced to nine months, the specified date for completion must be within seven months: s. 34A(5)).

4.2.4.3 Certificates of completion The Road Traffic Offenders Act 1988, s. 34B contains provisions of a mainly administrative nature in relation to the supply of certificates of completion. On the satisfactory completion of a course, the course organiser is obliged, within 14 days, to provide the offender with a certificate to that effect. The certificate must contain such particulars and be in such form as the Secretary of State may determine. Until this certificate is actually received by the clerk of the supervising court, the offender will not be regarded as having satisfied the terms of the s. 34A order and will not be entitled to take advantage of the reduced period of disqualification. If the certificate is received before expiration of the disqualification imposed under s. 34 but after the end of the period as would have been reduced by s. 34A, the offender will be entitled to a reduction in the period of disqualification from the date on which the certificate is actually received.

Example D is disqualified for 12 months and an order is made under s. 34A. If he completes a course satisfactorily, he will be entitled to have his disqualification reduced by three months. D completes a course and is issued with a certificate to that effect. The clerk of the court does not receive the certificate until 10 months after the initial date of disqualification. D is still disqualified from driving between months nine and 10. D is not disqualified from driving after month 10 because that is when the clerk of the court actually received the certificate. The net result is that D's reduced period of disqualification is two months and not three.

There are three grounds on which a course organiser can refuse to issue an offender with the necessary certificate, namely:

(a) if the offender fails to pay the course fees;
(b) if he fails to attend the course in accordance with the organiser's reasonable instructions; and
(c) if he fails to comply with any other reasonable requirements of the course organiser.

If a course organiser, for any of the above reasons, decides not to issue a certificate, he is required to give written notice of that decision to the offender as soon as possible, and in any event, not later than 14 days after the date specified in the order as being the latest date for completion of the course. So what will happen if a course organiser fails to give written notice to the offender as required? The answer is provided by s. 34B(7) of the 1988 Act. The offender may apply to the supervising court for a declaration that the organiser is in default of his obligation. If the court agrees and makes the declaration, s. 34A of the Act shall have effect as if the certificate had been received by the clerk of the court. The same remedy lies for an offender who is not issued with a certificate of completion because, for some other reason, the organiser has not issued a certificate. In other words, the court can 'issue its own certificate' if satisfied either that the offender has completed the course and not been issued with a certificate, or that the organiser has (albeit with good reason) not given an offender written notice of his decision not to issue a certificate.

Whenever a certificate is issued, or a court 'issues a certificate' in the circumstances outlined above, the Secretary of State must be notified in such a manner as he may determine (s. 34B(9)).

4.2.4.4 Supplementary provisions to ss. 34A and 34B The Road Traffic Offenders Act 1988, s. 34C contains miscellaneous provisions mainly concerned with matters of interpretation. The interpretation provisions are self-explanatory, but the one provision worthy of mention here is contained in s. 34C(1) which provides that the Secretary of State may issue guidance to course organisers as to the conduct of courses to which they *must* have regard. Further, if an offender is refused a certificate on the grounds that he has failed to comply with the organiser's reasonable instructions, in determining whether any instructions were reasonable, the court *shall* have regard to guidance given to organisers by the Secretary of State.

4.2.4.5 Experimental period for s. 34A All the provisions relating to courses for drink-drive offenders are made subject to s. 31 of the 1991 Act, which provides for an experimental period during which time the success of the scheme will be reviewed and evaluated. The final date for the conclusion of the experimental period is the end of 1997, unless by that time the Secretary of State makes a new order extending the period. So the scheme is initially intended to run for some six years. The proviso to this, however, is that the Secretary of State is empowered by s. 31(8) to extend the period by making an order to that effect, although this can only be done on one occasion. There is no power to make repeat orders which continually extend the duration of the scheme. Throughout the period that the scheme is in operation, whether or not the scheme continues beyond 1997, that period is to be known as the 'experimental period'.

During the experimental period, offenders who are convicted of the new offence of causing death by careless driving while unfit (Road Traffic Act 1988, s. 3A), are excluded from the operation of s. 34A. The reason for this exclusion is not clear, but it seems that, initially, the government wants to target those offenders who commit simple drink-drive offences. This will make evaluation and appraisal of the scheme's success easier and more reliable. The drunken driver whose careless driving results in death is, perhaps, less likely to re-offend in terms of committing the same offence, and one of the aims of the scheme will be to reduce the reconviction rate for the same offence.

Further, during the experimental period, only courts designated by the Secretary of State will be able to make orders under the Road Traffic Offenders Act 1988, s. 34A. As yet, no information is available as to which courts will be so designated.

4.2.5 Disqualification until test passed

A new s. 36 is substituted into the Road Traffic Offenders Act 1988, by virtue of s. 32 of the 1991 Act. The new section places a duty on courts to disqualify certain offenders from driving until they have passed what the Act now calls an 'appropriate' driving test. An appropriate driving test will be either a standard Department of Transport driving test, or a new extended driving test depending upon the type of offence of which the offender has been convicted. In certain other situations, the courts will have a discretion as to whether or not to order an offender to undergo a standard retest.

4.2.5.1 Mandatory retesting Where an offender is disqualified under the Road Traffic Offenders Act 1988, s. 34, for any of the following offences, the court *must* order him to be disqualified until the appropriate driving test is passed. Those offences are:

(a) manslaughter by the driver of a motor vehicle,
(b) causing death by dangerous driving,
(c) dangerous driving.

By s. 36(3), mandatory retesting will also apply to those offenders disqualified under ss. 34 or 35 in such circumstances as the Secretary of State may by order prescribe, or to those offenders convicted of an offence involving obligatory endorsement as may be so prescribed. The potential therefore exists to extend the concept of mandatory retesting to a very wide range of offences.

While not specifically referred to in the Act, the Road Traffic Review wished to see all those offenders disqualified for 12 months or more compelled to take a retest. This would mean that virtually all drink-drive offenders would be disqualified until satisfactorily completing a retest. This, it was felt, would be an added deterrent to those who contemplate driving after drinking. The Act now empowers the Secretary of State, by order, to include such offenders within these provisions although, at the time of writing, no such order has been made. Indeed, the Secretary of State may, if he so wishes, make an order compelling courts to order the retesting of offenders disqualified under the Road Traffic Offenders Act 1988, s. 35 (disqualification under the penalty points system).

4.2.5.2 Discretionary retesting In any other case which falls outside the scope of the Road Traffic Offenders Act 1988, s. 36(1), the court may, at its discretion, disqualify the offender until an appropriate driving test is passed, the only proviso being that the offence for which he has been convicted must be one that attracts a mandatory endorsement of his licence (s. 36(4)). In determining whether or not to make an order under this subsection, the court must have regard to the safety of road users (s. 36(6)).

4.2.5.3 The type of test There will be two types of retest to which an offender may be subjected. For the three specified offences (see above, 4.2.5.1) and for those offences which the Secretary of State names by order under s. 36(3) or where the offender is disqualified under s. 35, the retest will be 'an extended driving test'.

The extended driving test will differ from the usual 'L' test in several ways, both in content and length. The two major differences are, first, that the test will be of some one hour 15 minutes duration, more than double the length of the standard 'L' test. The rationale for this is that the longer the duration, the less likelihood of flaws being hidden from the examiner. Secondly, during the course of the test, the candidate will have to drive (where practicable) for an unspecified period of time on an unrestricted dual carriageway, thus making the test more appropriate for a driver with greater experience. Responsibility for the standard and reliability of assessment for the test will lie with the Department of Transport's driving examiners. The fee will be approximately double that of the standard test fee and the full cost must be borne by the candidate.

The main difficulty with retesting a 'dangerous' driver perhaps lies in the inability of a retest to detect irresponsible attitudes. A driver may well have considerable driving skills behind the wheel and, when on test, will drive accordingly. How can an examiner ever be sure that such a person will not revert to irresponsible and aggressive driving behaviour once the test is completed? The answer is he cannot. This possibility was recognised by the Review when making its recommendations, but it was nevertheless felt that an extended form of retest was the best that could be devised in the circumstances and was better than nothing at all.

In those cases where a court orders a retest in its discretion under s. 36(4), the retest will be the standard 'L' test of competence to drive as prescribed by the Road Traffic Act 1988, s. 89(3).

4.2.5.4 Other provisions Where a person is disqualified under the provisions of the Road Traffic Offenders Act 1988, s. 36 until he passes an extended driving test, any earlier order will cease to have effect and no further order shall be made while he is so disqualified (s. 36(7)). So, if an offender is ordered to take an ordinary retest for some offence and then re-offends so that the court is compelled to order an extended test, the earlier order is 'overruled' and substituted by the new order. Once this has occurred, no further order imposing a retest can be made. The disqualification ends on production to the Secretary of State of evidence in the prescribed form that the test has been passed (s. 36(8)).

The overall effect of s. 36 and its operation can be seen in the following example.

Example On 2 January 1992, D is disqualified for three months for speeding. The court exercises its discretion and further disqualifies him until he passes a retest. This will be the standard 'L' test of competence to drive (ss. 36(4) and (5)).

On 2 March 1992, D is convicted of dangerous driving. He is disqualified for 12 months and the court orders him (as it must) to be disqualified from driving until he passes an extended test of competence to drive (ss. 36(1) and (2)). The earlier order of disqualification until the standard test is passed, made on 2 January 1992, ceases to have effect and is superseded by this order (s. 36(7)).

On 2 March 1993, D is eligible to obtain a provisional licence, notwithstanding the disqualification imposed on 2 March 1992 (s. 37(3)). He must comply with the conditions of a provisional licence. If he drives and fails to comply with such conditions, he commits offences under Road Traffic Act 1988, ss. 87(1) and 103. He may, therefore, be arrested and dealt with for the offence of driving while disqualified.

On 2 May 1993, D successfully completes an extended driving test. His disqualification now ends on production of evidence of the passing of the test to the Secretary of State.

4.2.6 Procedure on short disqualifications

A minor amendment which provides for a new s. 37(A) to s. 37 of the Road Traffic Offenders Act 1988, is given effect by s. 33 of the 1991 Act. The effect of the amendment is to make clear that where an offender is disqualified for an obligatory endorseable offence and any disqualification imposed is for less than 56 days, or where an interim disqualification is imposed under the provisions of s. 26 of the 1988 Act, the licence will come back into force at the end of the period of disqualification without further action on the part of the offender.

4.3 Conditional Fixed Penalties

Section 34 of the 1991 Act substitutes new ss. 75, 76 and 77 into the Road Traffic Offenders Act 1988 (which applied only to Scotland) and provides for the issue of a conditional fixed penalty offer, the effect of the offer and payment of penalty and endorsement procedures to apply throughout England, Wales and Scotland. This standardisation of procedure throughout Great Britain is to be welcomed.

For the most part, the procedure is self-explanatory. The offences which attract the fixed penalty procedure are listed in the Road Traffic Offenders Act 1988, sch. 3 and sch. 5, and in practice will be most frequently invoked for speeding and parking offences. Where, in Scotland, a relevant offence has been committed, a constable or a procurator fiscal may issue or send a notice to the alleged offender (ss. 75(2) and (3)). The notice is referred to as a 'conditional offer' (s. 75(5)). The notice must contain a certain minimum amount of information, namely, particulars of the alleged offence, the amount of the fixed penalty and that proceedings cannot commence for at least 28 days or such longer period as the notice may specify (s. 75(7)). If the alleged offender pays the amount of the fixed penalty and delivers his driving licence to the fixed penalty clerk within the prescribed period, any liability to conviction for the offence is discharged.

As before, the fixed penalty system can only operate in this manner if the alleged offender is not liable to disqualification under the 'totting up' provisions of s. 35 of the 1988 Act. When deciding whether or not an offender is liable to be disqualified under s. 35, it shall be assumed that the number of penalty points to be attributed for the offence in question is the minimum possible (s. 75(9)). Should it be found that the alleged offender is liable to disqualification, then the procedure cannot be implemented and proceedings for the offence will be instituted in the normal way. In such cases, the fixed penalty clerk must notify the Chief Officer of Police or procurator fiscal, as appropriate, before such proceedings can be implemented (ss. 76(4) and (5)).

Assuming that the offender is not liable to be disqualified, his licence will be endorsed in the usual way and then returned to him. The endorsement must also be notified to the Secretary to State (s. 76(6)).

Chapter 5
Miscellaneous Provisions

Section 35 makes certain provisions relating to disabled persons' badges. Sections 36 and 37 amend the Powers of Criminal Courts Act 1973 and the Criminal Procedure (Scotland) Act 1975 to enable courts to order the forfeiture of vehicles used to commit certain road traffic offences. Sections 38 and 39 amend the same Acts to provide for disqualification where a vehicle is used for an assault or certain other offences. Section 40 empowers authorities to install and maintain equipment for the detection of traffic offences. Sections 41, 42, 43 and 44 deal with miscellaneous parking matters. Section 45 amends the Road Traffic Regulation Act 1984, enabling variable speed limits to be imposed by highway authorities. Section 46 amends the Road Traffic Regulation Act 1984 and the Road Traffic Act 1988 as they relate to tramcars and trolley vehicles. Section 47 relates to applications for licences to drive a hackney carriage. Sections 48 and 49 deal with minor procedural amendments.

5.1 Disabled Persons' Badges

5.1.1 Amendments to deter misuse of a disabled person's badge

How many times have you seen a vehicle displaying an orange disabled person's badge parked on double yellow lines in the centre of town, only to see the driver walking briskly back to the vehicle carrying several bags of shopping? A minority of people have abused the disabled persons' badge scheme, and this has now prompted a legislative response in the 1991 Act.

Section 35 of the 1991 Act makes two amendments to existing legislation. The first change is to the Chronically Sick and Disabled Persons Act 1970, s. 21. This section provides for badges to be displayed on motor vehicles used by disabled persons, and the major amendment comes in the form of an addition onto the existing s. 21(4). New s. 21(4A), (4B) and (4C) create an offence of displaying a disabled person's badge when driving a motor vehicle on a road, except where the badge is issued and displayed in accordance with the regulations.

The second change amends the Road Traffic Regulation Act 1984, s. 117, by replacing the existing s. 117(1) and (2) with new provisions to provide for regulations to prescribe the circumstances in which a badge may be lawfully displayed in order to benefit from a disabled person's parking concession. In short, misuse of the disabled persons' badge scheme is now an offence, and this, it is hoped, will deter those who are presently abusing the scheme.

5.1.2 Penalty for misuse of a disabled person's badge

A person guilty of this offence is liable on summary conviction to a fine not exceeding level 3 on the standard scale.

5.2 Forfeiture of Vehicles

The Power of Criminal Courts Act 1973, s. 43, or, in Scotland, the Criminal Procedure (Scotland) Act 1975, ss. 223 and 436, are frequently invoked in respect of motor vehicles which are used for purely criminal purposes, although it has never been entirely clear whether an offender can be deprived of his vehicle where he has committed only road traffic offences. Initially, the question was unlikely to arise, as the 1973 Act originally applied only to offences which rendered the offender liable to a term of imprisonment of two years or more. This obstacle was removed by the Criminal Justice Act 1988, s. 69, which extended the 1973 Act to all offences regardless of any minimum term of imprisonment. Even so, it has always been doubted whether vehicle confiscation for the simple road traffic offender was possible or indeed appropriate. These Acts are now amended to provide for the forfeiture of vehicles when certain road traffic offences are committed, thus all doubt is removed.

Section 36 of the 1991 Act inserts a new s. 43(1B) into the Powers of Criminal Courts Act 1973, which provides that where a person commits an offence to which that subsection applies, by driving, attempting to drive or being in charge of a vehicle, or failing to provide a breath specimen or failing to stop and give information or to report an accident, the vehicle *shall* be regarded as having been used for the purpose of committing the offence. New s. 43(1C) applies s. 43(1B) to any offence under the Road Traffic Act 1988 which is punishable with imprisonment, manslaughter (or culpable homicide in Scotland) and wanton and furious driving contrary to the Offences against the Person Act 1861, s. 35. So henceforth, drink-drive offenders, dangerous drivers, those who fail to stop after accidents, disqualified drivers etc. will find themselves liable to have their vehicles forfeited.

The general effect of s. 43 remains the same, however. An order requiring forfeiture should only be made in straightforward cases. The major area of difficulty has arisen where the defendant is either a joint owner of the property, or is not the owner at all. In such cases, no forfeiture order should be made (*R* v *Troth* [1980] RTR 389; *R* v *Maidstone Crown Court, ex parte Gill* [1986] 1 WLR 1405). However, this must be read subject to s. 43(4), which provides that where a person claims ownership of the property in question, that person may make application under the Police (Property) Act 1897 for the return of the property. Such application will not succeed if made later than six months from the date of the forfeiture order, and the applicant must satisfy the court that he did not consent to the offender having possession of the vehicle, or, if he did consent, that he did not know and had no reason to suspect that it would be used for criminal purposes. So, for example, if X owns a vehicle and he lends it to D, knowing that D is disqualified from driving, the vehicle may be forfeited. Similarly, if D is X's son and D is in the habit of borrowing his father's car on a Friday night in order to go on a pub crawl, and X knows this is D's habit, then the vehicle may well be forfeited, albeit that D is not the owner.

In considering whether or not to make a forfeiture order, the court must have regard to the value of the property and the likely financial and other effects upon the offender (s. 43(1A)). In effect, the court is being asked not to make a forfeiture order if its effect is to make the overall penalty out of proportion to the gravity of the offence (see *R* v *Scully* (1985) 7 Cr App R (S) 119).

5.3 Disqualification Where Vehicle is Used for Assault

Section 38 of the 1991 Act amends the Powers of Criminal Courts Act 1973, s. 44, by inserting a new subsection 1(A) into that section, giving any court the power to disqualify an offender who has used a motor vehicle to commit any form of assault. Before this amendment, the power to disqualify was restricted to the Crown Court where a person had been convicted of an offence which rendered him liable to be imprisoned for a term of two years or more. The power can now be exercised by the magistrates' court where the offence is one of an assault (including aiding, abetting, counselling or procuring, or inciting an offence) and where the court is satisfied that the assault was committed by the driving of a motor vehicle. In any case to which s. 44(1A) applies the court has the power to disqualify the offender for such period as it sees fit (s. 44(2A)).

Section 39 of the 1991 Act amends the Criminal Procedure (Scotland) Act 1975, ss. 223 and 436, by inserting ss. 223A and 436A which, essentially, achieve the same effect in respect of those offenders north of the border.

5.4 Other Miscellaneous Provisions

The remaining provisions contained in Part I of the Act 1991 vary in their importance, and for the most part, make amendments to existing legislation which have become necessary because of this Act. These amendments are largely self-explanatory, and reference may be made to the Act for precise details. The most important provisions in a practical sense are those amendments which relate to parking, both inside and outside London, the amendment relating to tramcars and trolley vehicles, the ability of local authorities to vary speed limits, and the new procedure to be adopted when a person applies for a licence to drive a hackney carriage.

5.4.1 Parking provisions

The Road Traffic Act 1991, sch. 3, enacted by s. 43 of the Act, provides for 'permitted' and 'special' parking areas outside London. The parking provisions are very similar to those contained in Part II of the Act (see below, 6.3). Briefly, they provide for county councils and metropolitan district councils in England, and county and district councils in Wales, to apply to the Secretary of State for orders designating permitted parking areas or special parking areas. Joint applications may also be made for permitted parking area orders. After the necessary consultation, the Secretary of State may make the orders requested. While such orders are in force, certain specified offences will be decriminalised (see sch. 3 para. (4)). Schedule 3 also enables the Secretary of State to vary the list of offences which may be decriminalised. The only proviso to this is that the offence to be decriminalised must be one which relates to a stationary vehicle. Further provisions also enable the Secretary of State, when making a parking area order, to apply such provisions of the new parking regime covered by Part II of the Act as he considers appropriate, with or without modification. So, for example, local authority and police wheelclamping could be extended nationwide.

The main purpose of these changes is to give local authorities much wider responsibilities for parking matters within their own areas. It will be for local authorities (rather than the police) to enforce breaches of parking regulations, and it is envisaged that the revenue collected from enforcement will cover the cost of administration. Sections 41 and 42 of the 1991 Act amend the Road Traffic Regulation Act 1984, ss. 35 and 46, enabling parking authorities to vary the charges imposed for parking either off-street or at a designated parking place.

5.4.2 Variable speed limits

Section 45 of the 1991 Act amends the Road Traffic Regulation Act 1984, s. 84, to enable local authorities to vary speed limits within their areas. This provision has attracted a good deal of publicity, with the Secretary of State appearing in the media and urging local authorities to use this new power to impose a 20 mph speed restriction on roads in the immediate vicinity of schools. This is in response to figures which show that the child road casualty rate is at its greatest when children are on their way to or from school. The statistics also show that around 10 per cent of all accidents are caused, or contributed to, by excess speed.

Under the amended s. 84, local authorities are empowered to impose speed restrictions on particular local roads permanently or temporarily, or even during certain times of the day for example, 8.00am to 9.00am and 3.00pm to 4.00pm.

5.4.3 Provisions relating to tramcars and trolley vehicles

Section 46 of the 1991 Act amends the existing law as it relates to tramcars and trolley vehicles, by inserting new sections into the Road Traffic Regulation Act 1984 and the Road Traffic Act 1988. This amendment was tabled in the Lords to meet a growing concern that such vehicles would be 'exempt' from road traffic legislation. This, it was felt, was no longer appropriate in view of the fact that several authorities are now constructing rapid light transit systems in their cities, for example, Manchester and South Yorkshire. The concern arose because these transit systems (although they depend on a railway) are actually built into the existing road network and will travel alongside pedestrians and other road traffic.

The enabling provision contained in s. 46 will allow the Secretary of State to make regulations to provide that certain sections of the two relevant Acts will apply to tramcars and trolley vehicles. The main sections that the Secretary of State may, by regulation, extend to tramcars and trolley vehicles are as follows:

(a) *Road Traffic Regulation Act 1984*
Offences of exceeding speed limits (ss. 81–89).

(b) *Road Traffic Act 1988*
Motor racing on public ways (s. 12).
The new construction and use offences (ss. 40A–42).
Provisions relating to test certificates (ss. 47 and 48).
Provisions relating to maintenance and loading (ss. 68–73).
Driver licensing, driver fitness and related disqualifications (ss. 87–109).
Insurance requirements (ss. 143–162).
Requirement to stop etc. following road accident (ss. 168 and 170).

Corresponding provisions of the Road Traffic Offenders Act 1988, are also included amongst those which the Secretary of State may, by regulation, apply to these vehicles. 'Tramcar' and 'trolley vehicle' are defined in the Road Traffic Act 1988, s. 192.

5.4.4 Applications for a licence to drive a hackney carriage

The amendment contained in s. 47 of the 1991 Act is a straightforward response to something which has been a matter of concern for some time, namely, the ability of taxi drivers to obtain a licence from the local authority without any substantial checks being made on their antecedents. This is now remedied by the insertion into the Local Goverment (Miscellaneous Provisions) Act 1976 of a new s. 51(1A). This amendment now enables the local authority to send a copy of the application submitted to the chief officer of police for the area concerned with a request for observations. The chief officer is required to respond to the request from the local authority.

In short, applicants for a hackney carriage licence will be vetted by the police.

5.4.5 Other matters in Part I

Briefly, the following matters are dealt with in the remaining sections of Part I of the 1991 Act:

(a) Powers to install equipment for the detection of traffic offences (s. 40).
(b) Variation of charges at off-street parking places (s. 41).
(c) Variation of charges at parking places (s. 42).
(d) Enactment of sch. 3 (s. 43).
(e) Appointment of parking attendants (s. 44).

Schedule 4 to the 1991 Act is brought into force by s. 48, and reference has been made to the contents of this schedule as appropriate in the text. The final section in Part I of the Act is s. 49, which has the effect of repealing Parts II, III and IV of the Road Traffic (Consequential Provisions) Act 1988, sch. 2. As these Parts concerned enactments which were never brought into force anyway, the change is of no practical significance.

Chapter 6
Traffic in London

Part II of the Act is concerned primarily with improving traffic conditions in London. It provides for the Secretary of State to designate a network of priority routes and appoint a Traffic Director for London to coordinate their implementation and operation. Contraventions of orders related to designated parking places will no longer be criminal offences and provision is made for authorities to impose penalties (recoverable as civil debts) for such contraventions.

Introduction Much of the legislation is of minor significance, merely laying down administrative foundations which will enable the Secretary of State to implement certain policies in respect of traffic flow in the capital. Consequently, those whose work requires them to have a detailed knowledge of the administrative aspects of this legislation should refer to the relevant part of the Act for guidance. This chapter will give an overview of the administrative sections contained in Part II, with more detailed consideration being given to those sections of greater importance.

6.1 Priority Routes

6.1.1 Designation of priority routes

The Road Traffic Act 1991, ss. 50 to 53 make provision for the designation of priority routes, traffic management guidance, the appointment of a traffic director and the director's network plan.

By order, the Secretary of State may designate any road in London a priority route. In exercising this power, the Secretary shall provide a network of priority routes, the aim of which is to improve the movement of traffic in the Metropolis. Before making any such order, he (the Secretary of State) is obliged to consult three particular bodies:

(a) the London authority through which the proposed priority route runs (s. 50(3)(a));
(b) the relevant Commissioner or Commissioners (s. 50(3)(b));
(c) London Regional Transport (s. 50(3)(c)).

Further, if the proposed priority route is likely to affect another road within the area of a London authority other than that already consulted, or a county council, obligatory consultation must take place with the authority or council so affected before the order is made (s. 50(4)).

The Secretary is required to issue the London authorities and the Director (as to whom, see 6.1.2 below) with 'guidance' with respect to the management of traffic in London, and in particular with respect to priority routes and the priority network (s. 51(1)). The guidance so issued may include provision setting out the Secretary's objectives in designating priority routes and with respect to the role of the Director. Such guidance as is issued may be varied at any time (s. 51(2)). Before issuing or varying such guidance, the Secretary *must* consult:

(a) such associations of London authorities as thought fit (s. 51(3)(a));
(b) the two Commissioners (s. 51(3)(b));
(c) the Disabled Persons Transport Advisory Committee (s. 51(3)(c));
(d) London Regional Transport (s. 51(3)(d)).

In preparing such guidance, the Secretary is required to have regard to the needs of people with a disability (s. 51(4)).

6.1.2 The Traffic Director

It shall be the duty of the Secretary of State to appoint a person to be known as the Traffic Director for London ('the Director'), and sch. 5 of the Act shall have effect with respect to the Director (s. 52(1) and (2)). Schedule 5 provides for such matters as the Director's status, tenure of office, remuneration, staff, accounts and annual report. In addition to specific duties imposed on the Director by legislation, he is required to have the general duty of coordinating the traffic management measures in relation to priority routes and to monitor those measures (s. 52(4)). As soon as is reasonably practicable after receiving the Secretary's traffic management guidance, the Director must submit to him (the Secretary), and to the London authorities, his plans for both the design and the operation of the priority route network, having regard to the traffic management guidance given by the Secretary and also having regard to the needs of people with a disability. This will be known as 'the network plan' (s. 53).

Before the network plan is submitted, the Director is obliged to consult with:

(a) the Secretary of State;
(b) the two Commissioners, as appropriate;
(c) any London authorities likely to be affected;
(d) county councils, as appropriate;
(e) associations of London authorities, as appropriate;
(f) London Regional Transport (s. 53(4)(a–f)).

The Director is empowered to vary the network plan, but must consult the aforementioned groups before doing so (s. 53(6)).

6.2 Local Plans

6.2.1 Preparation of local plans

Sections 54 to 62 of the 1991 Act make provisions which relate to local plans and trunk road local plans. After receiving the Secretary of State's traffic management guidance and the Director's network plan (see 6.1.2), each London authority is required to prepare a local plan for the operation of the priority routes within its area and for which it is the highway authority. Any local plan has to be prepared and submitted in accordance with the timetable which is set out in the network plan. In the preparation of the local plan, the authority is directed to have regard to the Secretary's guidance and the network plan (s. 54(6)).

The authority must indicate as part of the plan which of its powers under the Road Traffic Regulation Act 1984 and the Highways Act 1980, it proposes to exercise. In this respect it must indicate how the proposals relate to the needs of people with a disability (s. 54(7)). Essentially, this deals with matters such as access to premises, effects on local amenities, provision of adequate parking places etc. In similar vein, the authority is required to consult with certain bodies, such as the Commissioners, other authorities as appropriate, and organisations which represent the interests of people with a disability.

After preparation, the plan is to be submitted to the Director for his approval. The Director will not give his approval unless he is satisfied that the plan is consistent:

(a) with the Secretary's guidance and with the network plan;
(b) with the costing of the authority's proposals;
(c) with the local plan timetable (s. 54(10)).

Section 55 empowers the Secretary to order the Director to prepare local plans for the operation of priority routes which are trunk roads, and s. 56 requires the Secretary to do likewise for those trunk roads which are not covered elsewhere in the Act. In short, this enables the Secretary to bypass both the authority concerned and the Director if he considers it expedient to do so. However, in such a case, the Secretary is still required to consult all the bodies aforementioned.

6.2.2 Implementation of local plans

Where a local plan has been approved by the Director, the authority concerned is then under a duty to implement the plan as soon as is reasonably practicable and to continue to act in a manner which remains compatible with it (s. 57). In respect of a local trunk road plan prepared in accordance with s. 55 or s. 56, it is the duty of the Director to implement the plan as soon as is reasonably practicable (s. 58(1)). In order to implement a plan under s. 58, the Director is given authority to order London authorities to exercise such of their powers as he may specify. If they fail to do so, the Director is able to exercise those powers that the London authority may have exercised as though the authority had exercised them (s. 58(7)). Any reasonable administrative expenses incurred by the Director in exercising this power may be recovered from the authority concerned as though it were a civil debt (s. 58(11)). The Secretary of State is given identical powers to the Director in these circumstances (s. 58(12)).

6.2.3 Variation of local plans

A London authority may subsequently vary its plans, but before doing so it must obtain the written consent of the Director (s. 59(1)). The Director may issue a direction to a London authority requiring it to vary its local plan in such manner as he may specify (s. 59(2)). If the authority fails to vary the plan as the Director specifies, he may vary it himself (s. 59(5)). However, the Director must consult the relevant bodies before he takes such action, namely, the authority, the Commissioner(s), London Transport and any other authority affected (s. 59(6)).

6.2.4 Proposed action by an authority which is likely to affect a priority route

A London authority is prohibited from exercising any power under the Highways Act 1980 or the Road Traffic Regulation Act 1984 (see 6.2.1, above), in a way which is likely to affect a priority route without, in effect, obtaining the consent of the Director (s. 60(1) and (3)). This does not apply if the plan was originally approved by the Director, or if the Director himself or the Secretary of State requests or orders the action (s. 60(2)).

6.2.5 Intervention powers and failing to implement local plans

In the event that a London authority fails to prepare a local plan in accordance with the requirements of this Act, or if it fails to submit their local plan in accordance with those requirements, the Director may order it to do so. If the authority fails to comply with any such order, then the Director is empowered to prepare his own local plan (s. 61(1) and (2)). The Director has the same power in the event that he refuses to approve any local plan submitted to him by a London authority either on that or on any subsequent occasion. In other words, the authority can keep trying until the Director loses patience and prepares his own local plan (s. 61(3) and (5)). In the event that the Director submits his own local plan, he has a duty to undertake prior consultation with the various bodies mentioned throughout this chapter (s. 61(6)). Again, administrative expenses incurred as a result of the Director taking this action are recoverable as a civil debt.

In the event that a London authority has prepared a satisfactory local plan but for some reason cannot or will not implement it, the Director can order the authority to take such steps as are required to implement it in a satisfactory manner and in accordance with any timetable as he may draw up (s. 62(1)). This section would also cater for the situation where the authority has had a local plan imposed upon it, either by the Director or by the Secretary of State, and it cannot or will not implement that plan. If the authority acts in a manner which is incompatible with its local plan (either its own or an imposed local plan), then the Director may order it to take such steps as he considers appropriate to remove, as far as is practicable, the effects of the action (s. 62(2)). Should an authority fail to comply with the order of the Director, then, subject to the Secretary of State agreeing, the Director may take such steps as are necessary to carry out the terms of his order to the authority, and in this respect the Director shall have all the powers which the London authority would have in connection with the implementation of its local plan (s. 62(3) and (5)). The Secretary of State may limit the exercise of the Director's powers in this respect, to the implementation of either the whole or any part of the local plan (s. 62(4)). As before, administrative expenses incurred are recoverable from the authority as a civil debt (s. 62(9)).

By now, the reader may have formed the impression that the London authorities are, in effect, presented with Hobson's Choice. The final say about any of the local plans and priority routes lies firmly in the hands of the Director and, ultimately, the Secretary of State. In the final analysis, the London authorities can have their wishes overridden by the Executive, and at their own expense. The only redress for a disgruntled authority would appear to be judicial review in the event of administrative non-compliance with the procedures.

6.3 Parking in London

Apparently there *are* parking spaces to be had somewhere in London, although many of us could probably raise strong arguments to the contrary. The remainder of the 1991 Act deals with parking in London, the decriminalisation of certain parking related offences, wheelclamping and appeals against penalties imposed for infringement of parking restrictions.

Certainly, there are few things more annoying than the motorist who unilaterally decides that parking restrictions do not apply to him. A motorist who inconsiderately parks his vehicle so as to cause a reduction in the size of the carriageway from two lanes to one, does much to impede the free flow of traffic for everyone. A stop of some two minutes in such circumstances can mean a delay of several minutes to others. Such behaviour can be infuriating for other motorists. At peak periods, even someone who stops merely to let out a passenger causes delay. If the 'priority route system' is to be effective, there must be adequate control of irresponsible parking. The following sections seek, *inter alia*, to achieve that result.

At the outset, it should be pointed out that while what follows relates specifically to London, sch. 3 of the Act enables the Secretary of State, on application, to extend these provisions to the rest of the country (see above, 5.4.1).

6.3.1 Parking guidance

Section 63(1) of the 1991 Act provides for the Secretary of State to issue guidance to London authorities with a view to those London authorities coordinating their action with respect to parking in the London area. The joint planning committee for London (established under the Local Government Act 1985, s. 5) is charged with the duty of making proposals to the Secretary of State as to the content of his guidance and of keeping that guidance under review (s. 63(2)(a) and (b)).

Again, strong emphasis is placed on the notion of consultation, and before the Secretary of State can issue any guidance, he is obliged to consult the Commissioner(s), London Regional Transport, the Disabled Persons Transport Advisory Committee, such associations of London authorities as thought fit and any other persons thought to be appropriate (s. 63(3)). It is a requirement of the Act that those concerned with the preparation of the Secretary of State's guidance are obliged to consider the needs of people with a disability (s. 63(4)).

So what kinds of things will the guidance seek to address? The only things the Act refers to specifically are financial matters. Section 63(5) states that the guidance may include provision with respect to appropriate levels for:

(a) parking charges;
(b) penalty charges;
(c) charges for removal, storage and disposal of vehicles; and
(d) charges for removal of the dreaded wheelclamp.

The guidance may be varied at any time by the Secretary of State (s. 63(6)).

6.3.2 Charges at designated parking places

Section 64 amends the Road Traffic Regulation Act 1984, by inserting a new s. 46(1A). The overall effect of this amendment is to provide for parking charges in Greater London to differ from those charged outside the area.

6.3.3 Decriminalisation of orders relating to London parking places

The Road Traffic Regulation Act 1984, ss. 47, 8 and 11 are amended by s. 65 of the 1991 Act in the following way. First, s. 47 (which creates offences relating to designated parking places) is disapplied as it relates to designated parking places in Greater London. Secondly, a new s. 8(1A) is introduced, the effect of which is to disapply s. 8(1) as it relates to orders made under s. 6 of the 1984 Act so far as it designates any parking places. Thirdly, s. 11 is amended to provide that contravention of an experimental traffic order is not an offence so far as the order designates any parking places in Greater London.

6.3.4 Parking penalties

Of course, the decriminalisation of certain parking offences does not mean that there will not be a penalty charge to pay for infringing the regulations. It simply means that enforcement and control will be a matter for the London authorities rather than a criminal matter to be dealt with by the police and the courts. The main section in the 1991 Act which deals with parking penalties is s. 66.

Section 66(1) provides that where a parking attendant has reason to believe that a penalty charge is payable in respect of a stationary vehicle in a designated parking place, he may:

(a) fix a penalty charge notice to the vehicle; or
(b) give such a notice to whoever appears to be in charge of the vehicle.

There are four circumstances when a penalty charge will become payable, namely:

(a) the vehicle has been left otherwise than as authorised by the order relating to the designated parking place;
(b) the vehicle has been left beyond the period of parking which has been paid for;
(c) no parking charge with respect to the vehicle has been paid;
(d) there has been a contravention of, or failure to comply with, any provision made by or under any order relating to the designated parking place (s. 66(2)).

The penalty ticket issued must contain a certain minimum amount of information, and some of this information is so obvious that one would think it would not need stating (for example, the ticket must contain the amount of the charge (s. 66(3)(b)) and the address to which payment should be sent (s. 66(3)(f))! The other information which must be stated on the ticket is:

(a) the ground(s) on which the ticket was issued;
(b) that the charge must be paid within 28 days;
(c) that if it is paid within 14 days the amount of the charge will be reduced by a 'specified proportion'; and
(d) that, if the penalty charge is not paid within 28 days, a 'notice to owner' may be served by the London authority on the person appearing to the authority to be the owner of the vehicle.

The 'specified proportion' referred to in (c) means such proportion as will be determined by the London authorities (s. 66(4)). While the Act was progressing through Parliament, a figure mooted was 25 per cent. Indeed, this figure was originally included in the Bill, but withdrawn at a later stage. It does seem, however, that the amount of the reduction will be in the region of 25 per cent.

It is an offence for a person other than the owner or person in charge of the vehicle to interfere with or remove a penalty charge notice which is fixed to it. An offender, on summary conviction, is liable to a fine not exceeding level 2 on the standard scale (s. 66(5) and s. 66(6)).

6.3.5 Recovery, disposal and immobilisation of vehicles

The general power to remove vehicles is contained in the Road Traffic Regulation Act 1984, s. 99, and this section remains unaltered. Section 67 of the 1991 Act, however, amends s. 101 of the 1984 Act to cater for the disposal of vehicles removed within the Greater London area. For vehicles recovered outside Greater London, the existing law remains unchanged.

New s. 101(4A) and (5A) are inserted into the 1984 Act, and their effect is as follows:

(a) If, before a vehicle found in Greater London is disposed of, the vehicle is claimed by a person who satisfies the authority that he is the owner, and who pays the penalty charge due and the costs of removal and storage, then the authority may permit that person to remove the vehicle (s. 101(4A)).

(b) If the vehicle has already been disposed of when the owner attempts to claim it, then providing less than one year has elapsed from the date of disposal, the authority is required to pay him the proceeds of sale (if any) less an amount to cover the penalty charge, the costs of storage and removal and the costs of disposal (s. 101(5A))

The authority to charge for removal, storage and disposal of vehicles is preserved in the Road Traffic Regulations Act 1984, s. 102, which receives technical amendments to cater for the separation of Greater London from the rest of the country (s. 68).

6.3.6 Vehicle immobilisation

The effect of s. 69(1) of the 1991 Act is to enable parking attendants to clamp stationary vehicles parked in designated parking places in any of the circumstances referred to in s. 66(2)(a), (b) or (c). A notice must be displayed on the clamped vehicle which:

(a) indicates the device has been fixed and warning the driver not to attempt to move the vehicle; and

(b) specifying the steps necessary to gain release (s. 69(2)).

The device may be removed only by or under the direction of a person who has been duly authorised by the authority to act for that purpose (s. 69(3)). The clamp will be removed only when both the fee for removal and the penalty charge have been paid. Removal or interference with the warning notice is an offence which may be punished on summary conviction with a fine not exceeding level 2 on the standard scale (s. 69(5) and (6)). Unauthorised removal of the clamp or an attempt thereat is also an offence punishable on summary conviction by a fine not exceeding level 3 on the standard scale.

6.3.7 Exemptions from s. 69

The provisions of s. 69 of the 1991 Act will not apply:

(a) to a vehicle which is displaying a current disabled person's badge;

(b) for the first 15 minutes by which the vehicle in question exceeds the permitted period of waiting, provided the appropriate charge was paid at the time of parking;

(c) where not more than 15 minutes have elapsed since the end of any unexpired time which is available at the relevant parking meter at the time of parking (s. 70(1)).

If s. 69 does not apply because of one of these exemptions, and in fact the vehicle was not being used by a person entitled to a disabled person's concession, the person using the vehicle commits an offence punishable on summary conviction by a fine not exceeding level 3 on the standard scale (s. 70(2)).

6.4 Procedural Matters in Schedule 6

This schedule is brought into effect by s. 66(7) of the 1991 Act, and deals with miscellaneous procedural matters concerning parking penalties. The broad effect of sch. 6 is as follows.

Where a penalty charge is not paid within the relevant 28-day period (see 6.3.4), a notice may be served on the owner. The notice will state that failure to pay within a further 28-day period may lead to an increased charge being payable. It will also state that the person on whom the notice is served may make representations against the notice in the prescribed manner. It must also state, in simple language, the effect of para. 5 of the schedule (see below, 6.4.2).

The recipient of a notice may make representations against the notice to the London authority concerned in such form as may be prescribed. For a representation to be heard, it must be made within the 28-day period, otherwise the authority may choose to disregard it.

6.4.1 The grounds for making representations

The grounds for making representations against a penalty notice are specified in sch. 6, para. 2(4) and these are the only grounds on which representations may be made, namely:

(a) the recipient has never owned the vehicle;

(b) the recipient ceased to be the vehicle owner before the date of the alleged contravention;

(c) the recipient became the owner after the date of the alleged contravention;

(d) the alleged contravention did not occur;

(e) the person who was in control of the vehicle allowed it to remain in the parking place without the consent of the owner;

(f) the designation order is invalid;

(g) the recipient is a vehicle hire firm and the vehicle in question was hired out under a hire agreement, and the person in charge of it signed a statement acknowledging his liability for such penalty notices during the hire period;

(h) the penalty charge exceeded the amount applicable in the circumstances.

It shall be the duty of the authority to consider representations made to it and, following due consideration, to serve the representor with notice of its decision. If the authority accepts the representations made the notice will be cancelled. If it does not accept the representations made, then the authority may issue a notice of rejection which will inform the representor that a charge certificate will be issued unless the penalty is paid within 28 days of the rejection notice being served. The rejection notice will also state that that the representor may appeal to the parking adjudicator, in which case service of the charge certificate will be suspended pending appeal, indicate the power of an adjudicator to award costs against an appellant and describe the general form and manner in which an appeal should be made.

6.4.2 *Parking adjudication*

Appeals may be made to the parking adjudicator within 28 days of the service of the notice of rejection, although the parking adjudicator has discretion to extend that period as he sees fit. He may hear representations on any of the points outlined above, including those not previously advanced by the representor to the authority. The adjudicator may then give the authority such directions as he sees fit, and the authority must comply forthwith (sch. 6, para. 5).

6.4.3 *Charge certificates*

Where a charge certificate is issued, either on failure of an appeal or otherwise, its effect is to increase the penalty charge notice by 50 per cent. The person on whom the certificate is served then has 14 days in which to pay the increased charge, and if he does so, the certificate is cancelled. If the charge is not paid, however, the authority may apply to the county court to enforce the charge. If the court so orders, the increased charge will be treated as if it were payable under a county court order.

6.4.4 *Invalid notices*

This applies where a county court orders payment of an increased charge *and* the person against whom the order is made makes a statutory declaration *and* the declaration is served on the court within 21 days of the date on which the original court order was served on him.

The declaration must state that the person making it did not receive the 'notice to owner' in question, or made representations but did not receive a 'rejection notice' or appealed to the adjudicator but had no response to the appeal. Where the statutory declaration is so served, the order of the court shall be deemed to be revoked, the charge certificate cancelled, and the district judge shall serve on the person making the declaration written notice indicating the result of the declaration having been served (sch. 6, para. 8(5)).

Following this procedure the authority may then begin the process again by serving a 'notice to owner' or referring the matter to the adjudicator etc., as appropriate.

6.4.5 *Offence to give false information*

Schedule 6, para. 9 provides that it shall be an offence, in response to a 'notice to owner', to recklessly or knowingly make representations which are false in a material particular. The offence is clearly regarded as serious as it is punishable with a fine not exceeding level 5 (the maximum) on the standard scale.

6.4.6 Service of documents
Schedule 6, para. 10 provides that service of the documents referred to in the schedule (i.e. 'notice to owner', 'notice of rejection' or 'charge certificate') may be effected by post.

6.5 Appeals and Miscellaneous Matters

6.5.1 Right to make representations
Section 71 of the 1991 Act provides that an owner, or person in charge of a vehicle, who recovers it from the authority in accordance with the amended Road Traffic Regulation Act 1984, s. 101, or who secures its release from clamping in accordance with s. 69 of this Act, must be informed of his right to make representations to the relevant authority and of the effect of s. 72 (appeals to parking adjudicator against a decision under s. 71, see 6.5.2). This information must be given in writing (s. 71(2)). There are five grounds upon which representations may be made, namely:

(a) there were no reasonable grounds to believe the vehicle had been left in any of the circumstances specified in s. 66(2)(a), (b) or (c);
(b) the vehicle was permitted to remain at rest in the parking space by someone who was in control without the consent of the owner;
(c) the vehicle was not in a designated parking place;
(d) the vehicle had been improperly clamped because a s. 70(1) exemption applied;
(e) the penalty or charge exceeded the amount applicable in the circumstances.

The representations must be made within 28 days of the representor being informed of his right so to do, otherwise the authority can choose to disregard them (s. 71(5)). The representations must be considered to a conclusion within 56 days and the authority must serve notice of the decision on the representor. If the authority accepts the representations made, it shall refund any sums paid by the representor under the 1984 Act, s. 101, or s. 69 of this Act, except any sums which were properly paid or deducted (s. 71(7)).

If the authority does not accept the representations made, the notice of this decision shall inform the representor of his right to appeal to the parking adjudicator, the power of the parking adjudicator to award costs against any person appealing to him, and describe the form and manner in which the appeal should be made (s. 71(8)).

Should the authority fail to respond to a representation within the required 56-day period, the representor is deemed to be successful in his application (s. 71(9)).

It is an offence to recklessly or knowingly make false representations. On conviction a person guilty of this offence is liable to a fine not exceeding level 5 on the standard scale (s. 71(10) and (11)).

6.5.2 *Appeals to the parking adjudicator*

Where the authority rejects the representations made under s. 71, the representor has 28 days (or such period as the adjudicator may otherwise allow) in which to appeal to the parking adjudicator (s. 72(1)).

The adjudicator may consider both the original and any new representations which the representor may wish to submit. If the representor succeeds on appeal, the adjudicator must direct the authority to make the necessary refund (s. 72(2)).

6.5.3 *Immobilisation of vehicles in London by police*

The police currently get their powers to clamp vehicles from the Road Traffic Regulation Act 1984, ss. 104 to 106. Section 75 of the 1991 Act inserts a new s. 106A into the 1984 Act with the following effect. Provision is made to enable the Secretary of State to extend the clamping powers of the police throughout the Greater London area. In the event that the Secretary decides to make such an order, s. 106 of the 1984 Act (experimental period for immobilisation of vehicles) shall cease to have effect.

In effect, the consequence of these amendments will be to enable the Traffic Director to ask the Secretary of State to extend the clamping powers of the police to the whole of Greater London. After the necessary consultation has taken place, the Secretary may make such an order.

6.5.4 *Appointment of parking adjudicators, special parking areas and enforcement*

The responsibility for appointing parking adjudicators lies primarily with the Joint Committee of the London authorities. Appointments are made subject to the consent of the Lord Chancellor (s. 73(2)). The adjudicator must have a five-year general qualification within the meaning of the Courts and Legal Services Act 1990, s. 71, and he will be appointed for a term not to exceed five years. The only grounds for removing a parking adjudicator from office are misconduct or his being unfit to discharge his duties.

The Secretary of State is empowered to make regulations as to the procedure to be followed in relation to proceedings before adjudicators (s. 73(11)). Any person who is required to attend a hearing before an adjudicator, or who is required to produce a document, and fails without reasonable excuse to do so, is guilty of an offence punishable on summary conviction by a fine not exceeding level 2 on the standard scale (s. 73(14)). The parking adjudicator is required to submit an annual report on the discharge of his functions to the Joint Committee (s. 73(17)).

6.5.5 *Special parking areas*

By virtue of s. 76 of the 1991 Act, it is open to a London authority to make application to the Secretary of State for an order designating the whole or any part of that authority's area as a special parking area. Before making such an order, the Secretary is required to consult the Commissioner(s) (s. 76(2)).

If such order is granted, the effect will be to decriminalise:

(a) the general parking or waiting offences;
(b) parking of vehicles on verges, footpaths etc.;
(c) driving/parking on cycle tracks.

This list may be supplemented by other offences on further orders but only insofar as they relate to stationary vehicles (s. 76(4)).

Section 77 provides that the owners of vehicles who commit what would otherwise have been an offence save for s. 76 above, will be liable to pay a penalty charge. The procedures in sch. 6 (representations, appeals etc., see 6.4) apply to special parking areas.

6.5.6 Enforcement

The final major section in the 1991 Act is s. 78, which provides for enforcement of sums payable under Part II of the Act. The section enables, *inter alia*, the Lord Chancellor, by order, to make provision for warrants of execution to be executed by certificated bailiffs (s. 78(2)). The section goes on to set out the meaning of 'certificated bailiff' (s. 78(6)), and the consequences of using a bailiff who is not 'certificated' (s. 78(7)).

The remaining miscellaneous sections of Part II and Part III are self-explanatory.

Road Traffic Act 1991

PART I GENERAL

Driving offences

Section

1. Offences of dangerous driving.
2. Careless, and inconsiderate, driving.

Drink and drugs

3. Causing death by careless driving when under influence of drink or drugs.
4. Driving under influence of drink and drugs.

Motoring events

5. Disapplication of sections 1 to 3 of the Road Traffic Act 1988 for authorised motoring events.

Danger to road-users

6. Causing danger to road-users.

Cycling

7. Cycling offences.

Construction and use

8. Construction and use of vehicles.
9. Vehicle examiners.
10. Testing vehicles on roads.
11. Inspection of vehicles.
12. Power to prohibit driving of unfit vehicles.
13. Power to prohibit driving of overloaded vehicles.
14. Unfit and overloaded vehicles: offences.
15. Removal of prohibitions.
16. Supply of unroadworthy vehicles etc.

Licensing of drivers

17. Requirement of licence.
18. Physical fitness.
19. Effects of disqualification.

Insurance

20. Exception from requirement of third-party insurance.

Information

21. Information as to identity of driver etc.

Trial

22. Amendment of Schedule 1 to the Road Traffic Offenders Act 1988.
23. Speeding offences etc: admissibility of certain evidence.
24. Alternative verdicts.
25. Interim disqualification.

Penalties

26. Amendment of Schedule 2 to the Road Traffic Offenders Act 1988.
27. Penalty points to be attributed to offences.
28. Penalty points to be taken into account on conviction.
29. Disqualification for certain offences.
30. Courses for drink-drive offenders.
31. Experimental period for section 30.
32. Disqualification until test is passed.
33. Short periods of disqualification.
34. Conditional offer of fixed penalty.

Miscellaneous

35. Disabled persons' badges.
36. Forfeiture of vehicles.
37. Forfeiture of vehicles: Scotland.
38. Disqualification where vehicle used for assault.
39. Disqualification in Scotland where vehicle used to commit offence.
40. Power to install equipment for detection of traffic offences.
41. Variation of charges at off-street parking places.
42. Variation of charges at designated parking places.
43. Permitted and special parking areas outside London.
44. Parking attendants.
45. Variable speed limits.
46. Tramcars and trolley vehicles.
47. Application for licences to drive hackney carriages etc.
48. Minor and consequential amendments.
49. Omission of enactments not brought into force.

PART II TRAFFIC IN LONDON

Priority routes

50. Designation of priority routes.
51. The Secretary of State's traffic management guidance.
52. The Traffic Director for London.
53. The Director's network plan.

Local plans and trunk road local plans

Parking in London

Miscellaneous

PART III SUPPLEMENTARY

SCHEDULES

Schedule 1—Amendment of Schedule 1 to the Road Traffic Offenders Act 1988.
Schedule 2—Amendment of Schedule 2 to the Road Traffic Offenders Act 1988.
Schedule 3—Permitted and special parking areas outside London.
Schedule 4—Minor and consequential amendments.
Schedule 5—The Traffic Director for London.
Schedule 6—Parking penalties.
Schedule 7—Minor and consequential amendments in relation to London.
Schedule 8—Repeals.

Road Traffic Act 1991

1991 Chapter 40. An Act to amend the law about road traffic.

[25 July 1991]

BE IT ENACTED by the Queen's most Excellent Majesty, by and with the advice and consent of the Lords Spiritual and Temporal, and Commons, in this present Parliament assembled, and by the authority of the same, as follows:—

PART I

GENERAL

Driving offences

1. Offences of dangerous driving.

For sections 1 and 2 of the Road Traffic Act 1988 there shall be substituted—

1. "Causing death by dangerous driving.

A person who causes the death of another person by driving a mechanically propelled vehicle dangerously on a road or other public place is guilty of an offence.

2. Dangerous driving.

A person who drives a mechanically propelled vehicle dangerously on a road or other public place is guilty of an offence.

2A. Meaning of dangerous driving.

(1) For the purposes of sections 1 and 2 above a person is to be regarded as driving dangerously if (and, subject to subsection (2) below, only if)—

(a) the way he drives falls far below what would be expected of a competent and careful driver, and

(b) it would be obvious to a competent and careful driver that driving in that way would be dangerous.

(2) A person is also to be regarded as driving dangerously for the purposes of sections 1 and 2 above if it would be obvious to a competent and careful driver that driving the vehicle in its current state would be dangerous.

(3) In subsections (1) and (2) above "dangerous" refers to danger either of injury to any person or of serious damage to property; and in determining for the purposes of those subsections what would be expected of, or obvious to, a competent and careful driver in a particular case, regard shall be had not only to the circumstances of which he could be expected to be aware but

also to any circumstances shown to have been within the knowledge of the accused.

(4) In determining for the purposes of subsection (2) above the state of a vehicle, regard may be had to anything attached to or carried on or in it and to the manner in which it is attached or carried."

2. Careless, and inconsiderate driving.
For section 3 of the Road Traffic Act 1988 there shall be substituted:

3. "Careless, and inconsiderate, driving.
If a person drives a mechanically propelled vehicle on a road or other public place without due care and attention, or without reasonable consideration for other persons using the road or place, he is guilty of an offence."

Drink and drugs

3. Causing death by careless driving when under influence of drink or drugs.
Before section 4 of the Road Traffic Act 1988 there shall be inserted—

3A. "Causing death by careless driving when under influence of drink or drugs.

(1) If a person causes the death of another person by driving a mechanically propelled vehicle on a road or other public place without due care and attention, or without reasonable consideration for other persons using the road or place, and—

(a) he is, at the time when he is driving, unfit to drive through drink or drugs, or

(b) he has consumed so much alcohol that the proportion of it in his breath, blood or urine at that time exceeds the prescribed limit, or

(c) he is, within 18 hours after that time, required to provide a specimen in pursuance of section 7 of this Act, but without reasonable excuse fails to provide it,

he is guilty of an offence.

(2) For the purposes of this section a person shall be taken to be unfit to drive at any time when his ability to drive properly is impaired.

(3) Subsection (1)(b) and (c) above shall not apply in relation to a person driving a mechanically propelled vehicle other than a motor vehicle."

4. Driving under influence of drink or drugs.
In section 4 of the Road Traffic Act 1988, in subsections (1), (2) and (3) for the words "motor vehicle" there shall be substituted the words "mechanically propelled vehicle".

Motoring events

5. Disapplication of sections 1 to 3 of the Road Traffic Act 1988 for authorised motoring events.
After section 13 of the Road Traffic Act 1988 there shall be inserted—

13A. "Disapplication of sections 1 to 3 for authorised motoring events.

(1) A person shall not be guilty of an offence under sections 1, 2 or 3 of this Act by virtue of driving a vehicle in a public place other than a road if he

shows that he was driving in accordance with an authorisation for a motoring event given under regulations made by the Secretary of State.

(2) Regulations under this section may in particular—

(a) prescribe the persons by whom, and limit the circumstances in which and the places in respect of which, authorisations may be given under the regulations;

(b) specify conditions which must be included among those incorporated in authorisations;

(c) provide for authorisations to cease to have effect in prescribed circumstances;

(d) provide for the procedure to be followed, the particulars to be given, and the amount (or the persons who are to determine the amount) of any fees to be paid, in connection with applications for authorisations;

(e) make different provisions for different cases."

Danger to road-users

6. Causing danger to road-users.

Before section 23 of the Road Traffic Act 1988 there shall be inserted—

22A. "Causing danger to road-users.

(1) A person is guilty of an offence if he intentionally and without lawful authority or reasonable cause—

(a) causes anything to be on or over a road, or

(b) interferes with a motor vehicle, trailer or cycle, or

(c) interferes (directly or indirectly) with traffic equipment,

in such circumstances that it would be obvious to a reasonable person that to do so would be dangerous.

(2) In subsection (1) above "dangerous" refers to danger either of injury to any person while on or near a road, or of serious damage to property on or near a road; and in determining for the purposes of that subsection what would be obvious to a reasonable person in a particular case, regard shall be had not only to the circumstances of which he could be expected to be aware but also to any circumstances shown to have been within the knowledge of the accused.

(3) In subsection (1) above "traffic equipment" means—

(a) anything lawfully placed on or near a road by a highway authority;

(b) a traffic sign lawfully placed on or near a road by a person other than a highway authority;

(c) any fence, barrier or light lawfully placed on or near a road—

(i) in pursuance of section 174 of the Highways Act 1980, section 8 of the Public Utilities Street Works Act 1950 or section 65 of the New Roads and Street Works Act 1991 (which provide for guarding, lighting and signing in streets where works are undertaken), or

(ii) by a constable or a person acting under the instructions (whether general or specific) of a chief officer of police.

(4) For the purposes of subsection (3) above anything placed on or near a road shall unless the contrary is proved be deemed to have been lawfully placed there.

(5) In this section "road" does not include a footpath or bridleway.
(6) This section does not extend to Scotland."

Cycling

7. Cycling offences.
For section 28 of the Road Traffic Act 1988 there shall be substituted—

28. "Dangerous cycling.
(1) A person who rides a cycle on a road dangerously is guilty of an offence.
(2) For the purposes of subsection (1) above a person is to be regarded as riding dangerously if (and only if)—
(a) the way he rides falls far below what would be expected of a competent and careful cyclist, and
(b) it would be obvious to a competent and careful cyclist that riding in that way would be dangerous.
(3) In subsection (2) above "dangerous" refers to danger either of injury to any person or of serious damage to property; and in determining for the purposes of that subsection what would be obvious to a competent and careful cyclist in a particular case, regard shall be had not only to the circumstances of which he could be expected to be aware but also to any circumstances shown to have been within the knowledge of the accused."

Construction and use

8. Construction and use of vehicles.
(1) At the beginning of Part II of the Road Traffic Act 1988 there shall be inserted—

"Using vehicle in dangerous condition

40A. Using vehicle in dangerous condition etc.
A person is guilty of an offence if he uses, or causes or permits another to use, a motor vehicle or trailer on a road when—
(a) the condition of the motor vehicle or trailer, or of its accessories or equipment, or
(b) the purpose for which it is used, or
(c) the number of passengers carried by it, or the manner in which they are carried, or
(d) the weight, position or distribution of its load, or the manner in which it is secured,
is such that the use of the motor vehicle or trailer involves a danger of injury to any person."

(2) For section 42 of that Act there shall be substituted—

41A. "Breach of requirement as to brakes, steering-gear or tyres.
A person who—
(a) contravenes of fails to comply with a construction and use requirement as to brakes, steering-gear or tyres, or
(b) uses on a road a motor vehicle or trailer which does not comply with such a requirement, or causes or permits a motor vehicle or trailer to be so used,
is guilty of an offence.

41B. Breach of requirement as to weight: goods and passenger vehicles.

(1) A person who—

(a) contravenes or fails to comply with a construction and use requirement as to any description of weight applicable to—

(i) a goods vehicle, or

(ii) a motor vehicle or trailer adapted to carry more than eight passengers, or

(b) uses on a road a vehicle which does not comply with such a requirement, or causes or permits a vehicle to be so used,

is guilty of an offence.

(2) In any proceedings for an offence under this section in which there is alleged a contravention of or failure to comply with a construction and use requirement as to any description of weight applicable to a goods vehicle, it shall be a defence to prove either—

(a) that at the time when the vehicle was being used on the road—

(i) it was proceeding to a weighbridge which was the nearest available one to the place where the loading of the vehicle was completed for the purpose of being weighed, or

(ii) it was proceeding from a weighbridge after being weighed to the nearest point at which it was reasonably practicable to reduce the weight to the relevant limit, without causing an obstruction on any road, or

(b) in a case where the limit of that weight was not exceeded by more than 5 per cent.—

(i) that that limit was not exceeded at the time when the loading of the vehicle was originally completed, and

(ii) that since that time no person has made any addition to the load.

42. Breach of other construction and use requirements.

A person who—

(a) contravenes or fails to comply with any construction or use requirement other than one within section 41A(a) or 41B(1)(a) of this Act, or

(b) uses on a road a motor vehicle or trailer which does not comply with such a requirement, or causes or permits a motor vehicle or trailer to be so used,

is guilty of an offence."

9. Vehicle examiners.

(1) Section 7 of the Public Passenger Vehicles Act 1981 and section 68(1) and (2) of the Road Traffic Act 1988 (which provide for the appointment of certifying officers, public service vehicle examiners and examiners of goods vehicles) shall cease to have effect, and after section 66 of the Road Traffic Act 1988 there shall be inserted—

"*Vehicle examiners*

66A. Appointment of examiners.

(1) The Secretary of State shall appoint such examiners as he considers necessary for the purpose of carrying out the functions conferred on them by

this Part of this Act, the Public Passenger Vehicles Act 1981, the Transport Act 1968 and any other enactment.

(2) An examiner appointed under this section shall act under the general directions of the Secretary of State.

(3) In this Part of this Act "vehicle examiner" means an examiner appointed under this section."

(2) Any reference in any Act, or in any instrument made under any Act, to a certifying officer or public service vehicle examiner appointed under the Public Passenger Vehicles Act 1981 or to an examiner appointed under section 68(1) of the Road Traffic Act 1988 shall, so far as may be appropriate in consequence of the preceding provisions of this section, be construed as a reference to an examiner appointed under section 66A of the Road Traffic Act 1988.

10. Testing vehicles on roads.

(1) Section 67 of the Road Traffic Act 1988 (tests on roads to ascertain compliance with certain requirements) shall be amended as follows.

(2) In subsection (1) for paragraph (a) there shall be substituted—

"(a) ascertaining whether the following requirements, namely—

(i) the construction and use requirements, and

(ii) the requirement that the condition of the vehicle is not such that its use on a road would involve a danger of injury to any person,

are complied with as respects the vehicle;".

(3) For subsection (2) there shall be substituted—

"(2) For the purpose of testing a vehicle under this section the examiner—

(a) may require the driver to comply with his reasonable instructions, and

(b) may drive the vehicle."

(4) In subsection (4)(b) for "68(1)" there shall be substituted "66A".

(5) In subsection (4)(e) for the words "under instructions of the" there shall be substituted the words "on behalf of a".

11. Inspection of vehicles.

Subsections (1) to (2) of section 8 of the Public Passenger Vehicles Act 1981 (inspection of public service vehicles) shall cease to have effect, and for section 68 of the Road Traffic Act 1988 (inspection of goods vehicles) there shall be substituted—

"Inspection of public passenger vehicles and goods vehicles

68. Inspection of public passenger vehicles and goods vehicles.

(1) A vehicle examiner—

(a) may at any time, on production if so required of his authority, inspect any vehicle to which this section applies and for that purpose detain the vehicle during such time as is required for the inspection, and

(b) may at any time which is reasonable having regard to the circumstances of the case enter any premises on which he has reason to believe that such a vehicle is kept.

(2) The power conferred by subsection (1) above to inspect a vehicle includes power to test it and to drive it for the purpose of testing it.

(3) A person who intentionally obstructs an examiner in the exercise of his powers under subsection (1) above is guilty of an offence.

(4) A vehicle examiner or a constable in uniform may at any time require any person in charge of a vehicle to which this section applies and which is stationary on a road to proceed with the vehicle for the purpose of having it inspected under this section to any place where an inspection can be suitably carried out (not being more than five miles from the place where the requirement is made).

(5) A person in charge of a vehicle who refuses or neglects to comply with a requirement made under subsection (4) above is guilty of an offence.

(6) This section applies to—

(a) goods vehicles,

(b) public service vehicles, and

(c) motor vehicles which are not public service vehicles but are adapted to carry more than eight passengers;

but subsection (1)(b) above shall not apply in relation to vehicles within paragraph (c) above or in relation to vehicles used to carry passengers for hire or reward only under permits granted under section 19 or 22 of the Transport Act 1985 (use of vehicles by educational and other bodies or in providing community bus services)."

12. Power to prohibit driving of unfit vehicles.

Section 9 of the Public Passenger Vehicles Act 1981 (unfit public services vehicles) shall cease to have effect, and for section 69 of the Road Traffic Act 1988 (unfit goods vehicles) there shall be substituted—

"Prohibition of unfit vehicles

69. Power to prohibit driving of unfit vehicles.

(1) If on any inspection of a vehicle under section 41, 45, 49, 61, 67, 68 or 77 of this Act it appears to a vehicle examiner that owing to any defects in the vehicle it is, or is likely to become, unfit for service, he may prohibit the driving of the vehicle on a road—

(a) absolutely, or

(b) for one or more specified purposes, or

(c) except for one or more specified purposes.

(2) If on any inspection of a vehicle under any of the enactments mentioned in subsection (1) above it appears to an authorised constable that owing to any defects in the vehicle driving it (or driving it for any particular purpose or purposes or for any except one or more particular purposes) would involve a danger of injury to any person, he may prohibit the driving of the vehicle on a road—

(a) absolutely, or

(b) for one or more specified purposes, or

(c) except for one or more specified purposes.

(3) A prohibition under this section shall come into force as soon as the notice under subsection (6) below has been given if—

(a) it is imposed by an authorised constable, or

(b) in the opinion of the vehicle examiner imposing it the defects in the vehicle in question are such that driving it, or driving it for any purpose within the prohibition, would involve a danger of injury to any person.

(4) Except where subsection (3) applies, a prohibition under this section shall (unless previously removed under section 72 of this Act) come into force at such time not later than ten days from the date of the inspection as seems appropriate to the vehicle examiner imposing the prohibition, having regard to all the circumstances.

(5) A prohibition under this section shall continue in force until it is removed under section 72 of this Act.

(6) A person imposing a prohibition under this section shall forthwith give notice in writing of the prohibition to the person in charge of the vehicle at the time of the inspection—

(a) specifying the defects which occasioned the prohibition;

(b) stating whether the prohibition is on all driving of the vehicle or driving it for one or more specified purposes or driving it except for one or more specified purposes (and, where applicable, specifying the purpose or purposes in question);

and

(c) stating whether the prohibition is to come into force immediately or at the end of a specified period.

(7) Where a notice has been given under subsection (6) above, any vehicle examiner or authorised constable may grant an exemption in writing for the use of the vehicle in such manner, subject to such conditions and for such purpose as may be specified in the exemption.

(8) Where such a notice has been given, any vehicle examiner or authorised constable may by endorsement on the notice vary its terms and, in particular, alter the time at which the prohibition is to come into force or suspend it if it has come into force.

(9) In this section "authorised constable" means a constable authorised to act for the purposes of this section by or on behalf of a chief officer of police.

69A. Prohibitions conditional on inspection etc.

(1) Where it appears to the person imposing a prohibition under section 69 of this Act that the vehicle is adapted to carry more than eight passengers, or is a public service vehicle not so adapted, the prohibition may be imposed with a direction making it irremovable unless and until the vehicle has been inspected at an official PSV testing station within the meaning of the Public Passenger Vehicles Act 1981.

(2) Where it appears to that person that the vehicle is of a class to which regulations under section 49 of this Act apply, the prohibition may be imposed with a direction making it irremovable unless and until the vehicle has been inspected at an official testing station.

(3) Where it appears to that person that the vehicle is one to which section 47 of this Act applies, or would apply if the vehicle had been registered under the Vehicles (Excise) Act 1971 more than three years earlier, the prohibition may be imposed with a direction making it irremovable unless and until the vehicle has been inspected, and a test certificate issued, under section 45 of this Act.

(4) In any other case, the prohibition may be imposed with a direction making it irremovable unless and until the vehicle has been inspected in accordance with regulations under section 72 of this Act by a vehicle examiner or authorised constable (within the meaning of section 69 of this Act)."

13. Power to prohibit driving of overloaded vehicles.

(1) Section 70 of the Road Traffic Act 1988 shall be amended as follows.

(2) In subsection (1)—

(a) after the words "where a goods vehicle" there shall be inserted the words ", or a motor vehicle adapted to carry more than eight passengers,";

(b) for the words "a goods vehicle examiner" there shall be substituted the words "a vehicle examiner";

(c) at the end there shall be added the words "or that by reason of excessive overall weight or excessive axle weight on any axle driving the vehicle would involve a danger of injury to any person".

(3) In subsection (2)—

(a) for "69(2)" there shall be substituted "69(6)";

(b) after the words "that limit" there shall be inserted the words "or, as the case may be, so that it is no longer excessive".

(4) In subsection (4), for the words "a goods vehicle examiner" there shall be substituted the words "a vehicle examiner".

14. Unfit and overloaded vehicles: offences.

For section 71 of the Road Traffic Act 1988 there shall be substituted—

71. "Unfit and overloaded vehicles: offences.

(1) A person who—

(a) drives a vehicle in contravention of a prohibition under section 69 or 70 of this Act, or

(b) causes or permits a vehicle to be driven in contravention of such a prohibition, or

(c) fails to comply within a reasonable time with a direction under section 70(3) of this Act,

is guilty of an offence.

(2) The Secretary of State may by regulations provide for exceptions from subsection (1) above."

15. Removal of prohibitions.

(1) For section 72 of the Road Traffic Act 1988 there shall be substituted—

72. "Removal of prohibitions.

(1) Subject to the following provisions of this section, a prohibition under section 69 or 70 of this Act may be removed by any vehicle examiner or authorised constable if he is satisfied that the vehicle is fit for service.

(2) If the prohibition has been imposed with a direction under section 69A(1) or (2) of this Act, the prohibition shall not be removed unless and until the vehicle has been inspected in accordance with the direction.

(3) If the prohibition has been imposed with a direction under section 69A(3) of this Act, subsection (1) above shall not apply; but the prohibition

shall be removed, by such person as may be prescribed, if (and only if) any prescribed requirements relating to the inspection of the vehicle and the issue and production of a test certificate have been complied with.

(4) If the prohibition has been imposed with a direction under section 69A(4) of this Act, the prohibition shall not be removed unless and until any prescribed requirements relating to the inspection of the vehicle have been complied with.

(5) A person aggrieved by the refusal of a vehicle examiner or authorised constable to remove a prohibition may, within the prescribed time and in the prescribed manner, appeal to the Secretary of State.

(6) The Secretary of State may make such order on the appeal as he thinks fit.

(7) Where a vehicle examiner or authorised constable removes a prohibition, he must forthwith give notice of the removal to the owner of the vehicle.

(8) The Secretary of State may require the payment of fees, in accordance with prescribed scales and rates, for the inspection of a vehicle with a view to the removal of a prohibition; and—

(a) payment of fees may be required to be made in advance, and

(b) the Secretary of State must ensure that all the scales and rates prescribed for the purposes of this subsection are reasonably comparable with—

(i) in the case of goods vehicles, the fees charged by virtue of section 51(1)(h) in respect of periodic examination, and

(ii) in the case of other vehicles, the fees charged by virtue of section 46(c).

(9) The Secretary of State may make regulations for prescribing anything which may be prescribed under this section and for regulating the procedure, and fees payable, on appeals to him under subsection (5) above.

(10) In this section "authorised constable" means a constable authorised to act for the purposes of this section by or on behalf of a chief officer of police.

72A. Official testing stations.

The Secretary of State may provide and maintain stations (in this Part of this Act referred to as "official testing stations") where inspections of goods vehicles for the purposes of section 72 may be carried out and may provide and maintain the apparatus for carrying out such inspections."

16. Supply of unroadworthy vehicles etc.

(1) Section 75 of the Road Traffic Act 1988 (vehicles not to be sold in unroadworthy condition or altered so as to be unroadworthy) shall be amended as follows.

(2) In subsection (3), sub-paragraph (iii) of paragraph (a) shall be omitted and for paragraph (b) there shall be substituted—

"(b) it is in such a condition that its use on a road would involve a danger of injury to any person".

(3) In subsection (4), after the words "that condition" there shall be inserted—

"(a)",

and at the end there shall be added the words "or

(b) would involve a danger of injury to any person."

(4) In subsection (6), paragraph (c) shall be omitted.

(5) After subsection (6) there shall be inserted—

"(6A) Paragraph (b) of subsection (6) above shall not apply in relation to a person who, in the course of a trade or business—

(a) exposes a vehicle or trailer for sale, unless he also proves that he took all reasonable steps to ensure that any prospective purchaser would be aware that its use in its current condition on a road in Great Britain would be unlawful, or

(b) offers to sell a vehicle or trailer, unless he also proves that he took all reasonable steps to ensure that the person to whom the offer was made was aware of that fact."

Licensing of drivers

17. Requirement of licence.

(1) In subsection (1) of section 87 of the Road Traffic Act 1988 (offence for person to drive if he is not the holder of a licence) for the words "if he is not the holder of" there shall be substituted the words "otherwise than in accordance with".

(2) In subsection (2) of that section (offence to allow a person to drive if he does not hold a licence) for the words "if that other person is not the holder of a licence authorising him" there shall be substituted the words "otherwise than in accordance with a licence authorising that other person".

(3) Sections 97(7) and 98(5) of the Road Traffic Act 1988 (which make it an offence to fail to comply with the conditions of certain licences) shall cease to have effect.

18. Physical fitness.

(1) In section 92 of the Road Traffic Act 1988 (physical fitness of drivers) at the end there shall be added—

"(10) A person who holds a licence authorising him to drive a motor vehicle of any class and who drives a motor vehicle of that class on a road is guilty of an offence if the declaration included in accordance with subsection (1) above in the application on which the licence was granted was one which he knew to be false."

(2) In section 94 of that Act (provision of information about disabilities) after subsection (3) there shall be inserted—

"(3A) A person who holds a licence authorising him to drive a motor vehicle of any class and who drives a motor vehicle of that class on a road is guilty of an offence if at any earlier time while the licence was in force he was required by subsection (1) above to notify the Secretary of State but has failed without reasonable excuse to do so."

(3) After section 94 of that Act there shall be inserted—

94A. "Driving after refusal or revocation of licence.

(1) A person who drives a motor vehicle of any class on a road otherwise than in accordance with a licence authorising him to drive a motor vehicle of that class is guilty of an offence if—

(a) at any earlier time the Secretary of State has in accordance with section 92(3) of this Act refused to grant such a licence, or has under section 93(1) or (2) revoked such a licence, and

(b) he has not since that earlier time held such a licence.

(2) Section 88 of this Act shall apply in relation to subsection (1) above as it applies in relation to section 87."

19. Effects of disqualification.

For section 103 of the Road Traffic Act 1988 there shall be substituted—

"Effects of disqualification

103. Obtaining licence, or driving, while disqualified.

(1) A person is guilty of an offence if, while disqualified for holding or obtaining a licence, he—

(a) obtains a licence, or

(b) drives a motor vehicle on a road.

(2) A licence obtained by a person who is disqualified is of no effect (or, where the disqualification relates only to vehicles of a particular class, is of no effect in relation to vehicles of that class).

(3) A constable in uniform may arrest without warrant any person driving a motor vehicle on a road whom he has reasonable cause to suspect of being disqualified.

(4) Subsections (1) and (3) above do not apply in relation to disqualification by virtue of section 101 of this Act.

(5) Subsections (1)(b) and (3) above do not apply in relation to disqualification by virtue of section 102 of this Act.

(6) In the application of subsections (1) and (3) above to a person whose disqualification is limited to the driving of motor vehicles of a particular class by virtue of—

(a) section 102 or 117 of this Act, or

(b) subsection (9) of section 36 of the Road Traffic Offenders Act 1988 (disqualification until test is passed),

the references to disqualification for holding or obtaining a licence and driving motor vehicles are references to disqualification for holding or obtaining a licence to drive and driving motor vehicles of that class."

Insurance

20. Exception from requirement of third-party insurance.

(1) Section 144 of the Road Traffic Act 1988 shall be amended as follows.

(2) In subsection (1) (which removes the requirement for third-party insurance or security where £15,000 is kept deposited with the Accountant General of the Supreme Court) for "£15,000" there shall be substituted "£500,000".

(3) After subsection (1) there shall be inserted—

"(1A) The Secretary of State may by order made by statutory instrument substitute a greater sum for the sum for the time being specified in subsection (1) above.

(1B) No order shall be made under subsection (1A) above unless a draft of it has been laid before and approved by resolution of each House of Parliament."

Information

21. Information as to identity of driver etc.

For section 172 of the Road Traffic Act 1988 there shall be substituted—

172. "Duty to give information as to identity of driver etc in certain circumstances.

(1) This section applies—

(a) to any offence under the preceding provisions of this Act except—

(i) an offence under Part V, or

(ii) an offence under section 13, 16, 51(2), 61(4), 67(9), 68(4), 96 or 120,

and to an offence under section 178 of this Act,

(b) to any offence under sections 25, 26 or 27 of the Road Traffic Offenders Act 1988,

(c) to any offence against any other enactment relating to the use of vehicles on roads, except an offence under paragraph 8 of Schedule 1 to the Road Traffic (Driver Licensing and Information Systems) Act 1989, and

(d) to manslaughter, or in Scotland culpable homicide, by the driver of a motor vehicle.

(2) Where the driver of a vehicle is alleged to be guilty of an offence to which this section applies—

(a) the person keeping the vehicle shall give such information as to the identity of the driver as he may be required to give by or on behalf of a chief officer of police, and

(b) any other person shall if required as stated above give any information which it is in his power to give and may lead to identification of the driver.

(3) Subject to the following provisions, a person who fails to comply with a requirement under subsection (2) above shall be guilty of an offence.

(4) A person shall not be guilty of an offence by virtue of paragraph (a) of subsection (2) above if he shows that he did not know and could not with reasonable diligence have ascertained who the driver of the vehicle was.

(5) Where a body corporate is guilty of an offence under this section and the offence is proved to have been committed with the consent or connivance of, or to be attributable to neglect on the part of, a director, manager, secretary or other similar officer of the body corporate, or a person who was purporting to act in any such capacity, he, as well as the body corporate, is guilty of that offence and liable to be proceeded against and punished accordingly.

(6) Where the alleged offender is a body corporate, or in Scotland a partnership or an unincorporated association, or the proceedings are brought against him by virtue of subsection (5) above or subsection (11) below, subsection (4) above shall not apply unless, in addition to the matters there mentioned, the alleged offender shows that no record was kept of the persons who drove the vehicle and that the failure to keep a record was reasonable.

(7) A requirement under subsection (2) may be made by written notice served by post; and where it is so made—

(a) it shall have effect as a requirement to give the information within the period of 28 days beginning with the day on which the notice is served, and

(b) the person on whom the notice is served shall not be guilty of an offence under this section if he shows either that he gave the information as soon as reasonably practicable after the end of that period or that it has not been reasonably practicable for him to give it.

(8) Where the person on whom a notice under subsection (7) above is to be served is a body corporate, the notice is duly served if it is served on the secretary or clerk of that body.

(9) For the purposes of section 7 of the Interpretation Act 1978 as it applies for the purposes of this section the proper address of any person in relation to the service on him of a notice under subsection (7) above is—

(a) in the case of the secretary or clerk of a body corporate, that of the registered or principal office of that body or (if the body corporate is the registered keeper of the vehicle concerned) the registered address, and

(b) in any other case, his last known address at the time of service.

(10) In this section—

"registered address", in relation to the registered keeper of a vehicle, means the address recorded in the record kept under the Vehicles (Excise) Act 1971 with respect to that vehicle as being that person's address, and

"registered keeper", in relation to a vehicle, means the person in whose name the vehicle is registered under that Act;

and references to the driver of a vehicle include references to the rider of a cycle.

(11) Where, in Scotland, an offence under this section is committed by a partnership or by an unincorporated association other than a partnership and is proved to have been committed with the consent or connivance or in consequence of the negligence of a partner in the partnership or, as the case may be, a person concerned in the management or control of the association, he (as well as the partnership or association) shall be guilty of the offence."

Trial

22. Amendment of Schedule 1 to the Road Traffic Offenders Act 1988.
Schedule 1 to this Act, which amends Schedule 1 to the Road Traffic Offenders Act 1988 (procedural requirements applicable in relation to certain offences), shall have effect.

23. Speeding offences etc: admissibility of certain evidence.
For section 20 of the Road Traffic Offenders Act 1988 (admissibility of measurement of speed by radar) there shall be substituted—

20. "Speeding offences etc: admissibility of certain evidence.

(1) Evidence (which in Scotland shall be sufficient evidence) of a fact relevant to proceedings for an offence to which this section applies may be given by the production of—

(a) a record produced by a prescribed device, and

(b) (in the same or another document) a certificate as to the circumstances in which the record was produced signed by a constable or by a person authorised by or on behalf of the chief officer of police for the police area in which the offence is alleged to have been committed;
but subject to the following provisions of this section.

(2) This section applies to—

(a) an offence under section 16 of the Road Traffic Regulation Act 1984 consisting in the contravention of a restriction on the speed of vehicles imposed under section 14 of that Act;

(b) an offence under subsection (4) of section 17 of that Act consisting in the contravention of a restriction on the speed of vehicles imposed under that section;

(c) an offence under section 88(7) of that Act (temporary minimum speed limits);

(d) an offence under section 89(1) of that Act (speeding offences generally);

(e) an offence under section 36(1) of the Road Traffic Act 1988 consisting in the failure to comply with an indication given by a light signal that vehicular traffic is not to proceed.

(3) The Secretary of State may by order amend subsection (2) above by making additions to or deletions from the list of offences for the time being set out there; and an order under this subsection may make such transitional provision as appears to him to be necessary or expedient.

(4) A record produced or measurement made by a prescribed device shall not be admissible as evidence of a fact relevant to proceedings for an offence to which this section applies unless—

(a) the device is of a type approved by the Secretary of State, and

(b) any conditions subject to which the approval was given are satisfied.

(5) Any approval given by the Secretary of State for the purposes of this section may be given subject to conditions as to the purposes for which, and the manner and other circumstances in which, any device of the type concerned is to be used.

(6) In proceedings for an offence to which this section applies, evidence (which in Scotland shall be sufficient evidence)—

(a) of a measurement made by a device, or of the circumstances in which it was made, or

(b) that a device was of a type approved for the purposes of this section, or that any conditions subject to which an approval was given were satisfied,

may be given by the production of a document which is signed as mentioned in subsection (1) above and which, as the case may be, gives particulars of the measurement or of the circumstances in which it was made, or states that the device was of such a type or that, to the best of the knowledge and belief of the person making the statement, all such conditions were satisfied.

(7) For the purposes of this section a document purporting to be a record of the kind mentioned in subsection (1) above, or to be a certificate or other document signed as mentioned in that subsection or in subsection (6)

above, shall be deemed to be such a record, or to be so signed, unless the contrary is proved.

(8) Nothing in subsection (1) or (6) above makes a document admissible as evidence in proceedings for an offence unless a copy of it has, not less than seven days before the hearing or trial, been served on the person charged with the offence; and nothing in those subsections makes a document admissible as evidence of anything other than the matters shown on a record produced by a prescribed device if that person, not less than three days before the hearing or trial or within such further time as the court may in special circumstances allow, serves a notice on the prosecutor requiring attendance at the hearing or trial of the person who signed the document.

(9) In this section "prescribed device" means device of a description specified in an order made by the Secretary of State.

(10) The powers to make orders under subsections (3) and (9) above shall be exercisable by statutory instrument, which shall be subject to annulment in pursuance of a resolution of either House of Parliament."

24. Alternative verdicts.

For section 24 of the Road Traffic Offenders Act 1988 there shall be substituted—

24. "Alternative verdicts: general.

(1) Where—

(a) a person charged with an offence under a provision of the Road Traffic Act 1988 specified in the first column of the Table below (where the general nature of the offences is also indicated) is found not guilty of that offence, but

(b) the allegations in the indictment or information (or in Scotland complaint) amount to or include an allegation of an offence under one or more of the provisions specified in the corresponding entry in the second column, he may be convicted of that offence or of one or more of those offences.

Offence charged	*Alternative*
Section 1 (causing death by dangerous driving)	Section 2 (dangerous driving) Section 3 (careless, and inconsiderate, driving)
Section 2 (dangerous driving)	Section 3 (careless, and inconsiderate, driving)
Section 3A (causing death by careless driving when under influence of drink or drugs)	Section 3 (careless, and inconsiderate, driving) Section 4(1) (driving when unfit to drive through drink or drugs) Section 5(1)(a) (driving with excess alcohol in breath, blood or urine) Section 7(6) (failing to provide specimen)
Section 4(1) (driving or attempting to drive when unfit to drive through drink or drugs)	Section 4(2) (being in charge of a vehicle when unfit to drive through drink or drugs)

Offence charged	*Alternative*
Section 5(1)(a) (driving or attempting to drive with excess alcohol in breath, blood or urine)	Section 5(1)(b) (being in charge of a vehicle with excess alcohol in breath blood or urine)
Section 28 (dangerous cycling)	Section 29 (careless, and inconsiderate, cycling)

(2) Where the offence with which a person is charged is an offence under section 3A of the Road Traffic Act 1988, subsection (1) above shall not authorise his conviction of any offence of attempting to drive.

(3) Where a person is charged with having committed an offence under section 4(1) or 5(1)(a) of the Road Traffic Act 1988 by driving a vehicle, he may be convicted of having committed an offence under the provision in question by attempting to drive.

(4) Where by virtue of this section a person is convicted before the Crown Court of an offence triable only summarily, the court shall have the same powers and duties as a magistrates' court would have had on convicting him of that offence.

(5) Where, in Scotland, by virtue of this section a person is convicted under solemn procedure of an offence triable only summarily, the penalty imposed shall not exceed that which would have been competent on a conviction under summary procedure.

(6) This section has effect without prejudice to section 6(3) of the Criminal Law Act 1967 (alternative verdicts on trial on indictment), sections 61, 63, 64, 312 and 457A of the Criminal Procedure (Scotland) Act 1975 and section 23 of this Act."

25. Interim disqualification.

For section 26 of the Road Traffic Offenders Act 1988 (interim disqualification on committal for sentence in England and Wales) there shall be substituted—

26. "Interim disqualification.

(1) Where a magistrates' court—

(a) commits an offender to the Crown Court under subsection (1) of section 56 of the Criminal Justice Act 1967, or any enactment to which that section applies, or

(b) remits an offender to another magistrates' court under section 39 of the Magistrates' Court Act 1980,

to be dealt with for an offence involving obligatory or discretionary disqualification, it may order him to be disqualified until he has been dealt with in respect of the offence.

(2) Where a court in England and Wales—

(a) defers passing sentence on an offender under section 1 of the Powers of Criminal Courts Act 1973 in respect of an offence involving obligatory or discretionary disqualification, or

(b) adjourns after convicting an offender of such an offence but before dealing with him for the offence,

it may order the offender to be disqualified until he has been dealt with in respect of the offence.

(3) Where a court in Scotland—

(a) adjourns a case under section 179 or section 380 of the Criminal Procedure (Scotland) Act 1975 (for inquiries to be made or to determine the most suitable method of dealing with the offender);

(b) remands a person in custody or on bail under section 180 or section 381 of the Criminal Procedure (Scotland) Act 1975 (to enable a medical examination and report to be made);

(c) defers sentence under section 219 or section 432 of the Criminal Procedure (Scotland) Act 1975;

(d) remits a convicted person to the High Court for sentence under section 104 of the Criminal Procedure (Scotland) Act 1975,

in respect of an offence involving obligatory or discretionary disqualification, it may order the accused to be disqualified until he has been dealt with in respect of the offence.

(4) Subject to subsection (5) below, an order under this section shall cease to have effect at the end of the period of six months beginning with the day on which it is made, if it has not ceased to have effect before that time.

(5) In Scotland, where a person is disqualified under this section where section 219 or section 432 of the Criminal Procedure (Scotland) Act 1975 (deferred sentence) applies and the period of deferral exceeds 6 months, subsection (4) above shall not prevent the imposition under this section of any period of disqualification which does not exceed the period of deferral.

(6) Where a court orders a person to be disqualified under this section ("the first order"), no court shall make a further order under this section in respect of the same offence or any offence in respect of which an order could have been made under this section at the time the first order was made.

(7) Where a court makes an order under this section in respect of any person it must—

(a) require him to produce to the court any licence held by him and its counterpart, and

(b) retain the licence and counterpart until it deals with him or (as the case may be) cause them to be sent to the clerk of the court which is to deal with him.

(8) If the holder of the licence has not caused it and its counterpart to be delivered, or has not posted them, in accordance with section 7 of this Act and does not produce the licence and counterpart as required under subsection (7) above, then he is guilty of an offence.

(9) Subsection (8) above does not apply to a person who—

(a) satisfies the court that he has applied for a new licence and has not received it, or

(b) surrenders to the court a current receipt for his licence and its counterpart issued under section 56 of this Act, and produces the licence and counterpart to the court immediately on their return.

(10) Where a court makes an order under this section in respect of any person, sections 44(1) and 47(2) of this Act and section 109(3) of the Road Traffic Act 1988 (Northern Ireland drivers' licences) shall not apply in relation to the order, but—

(a) the court must send notice of the order to the Secretary of State, and

(b) if the court which deals with the offender determines not to order him to be disqualified under section 34 or 35 of this Act, it must send notice of the determination to the Secretary of State.

(11) A notice sent by a court to the Secretary of State in pursuance of subsection (10) above must be sent in such manner and to such address and contain such particulars as the Secretary of State may determine.

(12) Where on any occasion a court deals with an offender—

(a) for an offence in respect of which an order was made under this section, or

(b) for two or more offences in respect of any of which such an order was made,

any period of disqualification which is on that occasion imposed under section 34 or 35 of this Act shall be treated as reduced by any period during which he was disqualified by reason only of an order made under this section in respect of any of those offences.

(13) Any reference in this or any other Act (including any Act passed after this Act) to the length of a period of disqualification shall, unless the context otherwise requires, be construed as a reference to its length before any reduction under this section.

(14) In relation to licences which came into force before 1st June 1990, the references in this section to counterparts of licences shall be disregarded."

Penalties

26. Amendment of Schedule 2 to the Road Traffic Offenders Act 1988.
Schedule 2 to this Act, which amends Schedule 2 to the Road Traffic Offenders Act 1988 (prosecution and punishment of offences), shall have effect.

27. Penalty points to be attributed to offences.
For section 28 of the Road Traffic Offenders Act 1988 there shall be substituted—

28. "Penalty points to be attributed to an offence.

(1) Where a person is convicted of an offence involving obligatory endorsement, then, subject to the following provisions of this section, the number of penalty points to be attributed to the offence is—

(a) the number shown in relation to the offence in the last column of Part I or Part II of Schedule 2 to this Act, or

(b) where a range of numbers is shown, a number within that range.

(2) Where a person is convicted of an offence committed by aiding, abetting, counselling or procuring, or inciting to the commission of, an offence involving obligatory disqualification, then, subject to the following provisions of this section, the number of penalty points to be attributed to the offence is ten.

(3) Where both a range of numbers and a number followed by the words "(fixed penalty)" is shown in the last column of Part I of Schedule 2 to this Act in relation to an offence, that number is the number of penalty points to be attributed to the offence for the purposes of sections 57(5) and 77(5) of this Act; and, where only a range of numbers is shown there, the lowest number in the range is the number of penalty points to be attributed to the offence for those purposes.

(4) Where a person is convicted (whether on the same occasion or not) of two or more offences committed on the same occasion and involving obligatory endorsement, the total number of penalty points to be attributed to them is the number or highest number that would be attributed on a conviction of one of them (so that if the convictions are on different occasions the number of penalty points to be attributed to the offences on the later occasion or occasions shall be restricted accordingly).

(5) In a case where (apart from this subsection) subsection (4) above would apply to two or more offences, the court may if it thinks fit determine that that subsection shall not apply to the offences (or, where three or more offences are concerned, to any one or more of them).

(6) Where a court makes such a determination it shall state its reasons in open court and, if it is a magistrates' court, or in Scotland a court of summary jurisdiction, shall cause them to be entered in the register (in Scotland, record) of its proceedings.

(7) The Secretary of State may by order made by statutory instrument—

(a) alter a number or range of numbers shown in relation to an offence in the last column of Part I or Part II of Schedule 2 to this Act (by substituting one number or range for another, a number for a range, or a range for a number),

(b) where a range of numbers is shown in relation to an offence in the last column of Part I, add or delete a number together with the words "(fixed penalty)", and

(c) alter the number of penalty points shown in subsection (2) above;

and an order under this subsection may provide for different numbers or ranges of numbers to be shown in relation to the same offence committed in different circumstances.

(8) Where the Secretary of State exercises his power under subsection (7) above by substituting or adding a number which appears together with the words "(fixed penalty)", that number shall not exceed the lowest number in the range shown in the same entry.

(9) No order shall be made under subsection (7) above unless a draft of it has been laid before and approved by resolution of each House of Parliament."

28. Penalty points to be taken into account on conviction.

For section 29 of the Road Traffic Offenders Act 1988 there shall be substituted—

29. "Penalty points to be taken into account on conviction.

(1) Where a person is convicted of an offence involving obligatory endorsement, the penalty points to be taken into account on that occasion are (subject to subsection (2) below)—

(a) any that are to be attributed to the offence or offences of which he is convicted, disregarding any offence in respect of which an order under section 34 of this Act is made, and

(b) any that were on a previous occasion ordered to be endorsed on the counterpart of any licence held by him, unless the offender has since that occasion and before the conviction been disqualified under section 35 of this Act.

(2) If any of the offences was committed more than three years before another, the penalty points in respect of that offence shall not be added to those in respect of the other.

(3) In relation to licences which came into force before 1st June 1990, the reference in subsection (1) above to the counterpart of a licence shall be construed as a reference to the licence itself."

29. Disqualification for certain offences.

(1) Section 34 of the Road Traffic Offenders Act 1988 (disqualification for certain offences) shall be amended as follows.

(2) For subsection (2) there shall be substituted—

"(2) Where a person is convicted of an offence involving discretionary disqualification, and either—

(a) the penalty points to be taken into account on that occasion number fewer than twelve, or

(b) the offence is not one involving obligatory endorsement,

the court may order him to be disqualified for such period as the court thinks fit."

(3) In subsection (3) before paragraph (a) there shall be inserted—

"(aa) section 3A (causing death by careless driving when under the influence of drink or drugs),".

(4) For subsection (4) there shall be substituted—

"(4) Subject to subsection (3) above, subsection (1) above shall apply as if the reference to twelve months were a reference to two years—

(a) in relation to a person convicted of—

(i) manslaughter, or in Scotland culpable homicide, or

(ii) an offence under section 1 of the Road Traffic Act 1988 (causing death by dangerous driving), or

(iii) an offence under section 3A of that Act (causing death by careless driving while under the influence of drink or drugs), and

(b) in relation to a person on whom more than one disqualification for a fixed period of 56 days or more has been imposed within the three years immediately preceding the commission of the offence.

(4A) For the purposes of subsection (4)(b) above there shall be disregarded any disqualification imposed under section 26 of this Act or section 44 of the Powers of Criminal Courts Act 1973 or section 23A or 436A of the Criminal Procedure (Scotland) Act 1975 (offences committed by using vehicles) and any disqualification imposed in respect of an offence of stealing a motor vehicle, an offence under section 12 or 25 of the Theft Act 1968, an offence under section 178 of the Road Trafic Act 1988, or an attempt to commit such an offence."

30. Courses for drink-drive offenders.
After section 34 of the Road Traffic Offenders Act 1988 there shall be inserted—

34A. "Reduced disqualification period for attendance on courses.
(1) This section applies where—
(a) a person is convicted of an offence under section 3A (causing death by careless driving when under influence of drink or drugs), 4 (driving or being in charge when under influence of drink or drugs), 5 (driving or being in charge with excess alcohol) or 7 (failing to provide a specimen) of the Road Traffic Act 1988, and
(b) the court makes an order under section 34 of this Act disqualifying him for a period of not less than twelve months.
(2) Where this section applies, the court may make an order that the period of disqualification imposed under section 34 shall be reduced if, by a date specified in the order under this section, the offender satisfactorily completes a course approved by the Secretary of State for the purposes of this section and specified in the order.
(3) The reduction made by an order under this section in a period of disqualification imposed under section 34 shall be a period specified in the order of not less than three months and not more than one quarter of the unreduced period (and accordingly where the period imposed under section 34 is twelve months, the reduced period shall be nine months).
(4) The court shall not make an order under this section unless—
(a) it is satisfied that a place on the course specified in the order will be available for the offender,
(b) the offender appears to the court to be of or over the age of 17,
(c) the court has explained the effect of the order to the offender in ordinary language, and has informed him of the amount of the fees for the course and of the requirement that he must pay them before beginning the course, and
(d) the offender has agreed that the order should be made.
(5) the date specified in an order under this section as the latest date for completion of a course must be at least two months before the last day of the period of disqualification as reduced by the order.
(6) An order under this section shall name the petty sessions area (or in Scotland the sheriff court district or, where an order has been made under this section by a stipendiary magistrate, the commission area) in which the offender resides or will reside.

34B. Certificates of completion of courses.
(1) An offender shall be regarded for the purposes of section 34A of this Act as having completed a course satisfactorily if (and only if) a certificate that he has done so is received by the clerk of the supervising court before the end of the period of disqualification imposed under section 34.
(2) If the certificate referred to in subsection (1) above is received by the clerk of the supervising court before the end of the period of disqualification imposed under section 34 but after the end of the period as it would have been reduced by the order, the order shall have effect as if the reduced period ended with the day on which the certificate is received by the clerk.

(3) The certificate referred to in subsection (1) above shall be a certificate in such form, containing such particulars, and given by such person, as may be prescribed by, or determined in accordance with, regulations made by the Secretary of State.

(4) A course organiser shall give the certificate mentioned in subsection (1) above to the offender not later than fourteen days after the date specified in the order as the latest date for completion of the course, unless the offender fails to make due payment of the fees for the course, fails to attend the course in accordance with the organiser's reasonable instructions, or fails to comply with any other reasonable requirements of the organiser.

(5) Where a course organiser decides not to give the certificate mentioned in subsection (1) above, he shall give written notice of his decision to the offender as soon as possible, and in any event not later than fourteen days after the date specified in the order as the latest date for completion of the course.

(6) An offender to whom a notice is given under subsection (5) above may, within such period as may be prescribed by rules of court, apply to the supervising court for a declaration that the course organiser's decision not to give a certificate was contrary to subsection (4) above; and if the court grants the application section 34A of this Act shall have effect as if the certificate had been duly received by the clerk of the court.

(7) If fourteen days after the date specified in the order as the latest date for completion of the course the course organiser has given neither the certificate mentioned in subsection (1) above nor a notice under subsection (5) above, the offender may, within such period as may be prescribed by rules of court, apply to the supervising court for a declaration that the course organiser is in default; and if the court grants the application section 34A of this Act shall have effect as if the certificate had been duly received by the clerk of the court.

(8) A notice under subsection (5) above shall specify the ground on which it is given, and the Secretary of State may by regulations make provision as to the form of notices under that subsection and as to the circumstances in which they are to be treated as given.

(9) Where the clerk of a court receives a certificate of the kind referred to in subsection (1) above or a court grants an application under subsection (6) or (7) above, the clerk or court must send notice of that fact to the Secretary of State; and the notice must be sent in such manner and to such address, and must contain such particulars, as the Secretary of State may determine.

34C. Provisions supplementary to sections 34A and 34B.

(1) The Secretary of State may issue guidance to course organisers, or to any category of course organiser as to the conduct of courses approved for the purposes of section 34A of this Act; and—

(a) course organisers shall have regard to any guidance given to them under this subsection, and

(b) in determining for the purposes of section 34B(6) whether any instructions or requirements of an organiser were reasonable, a court shall have regard to any guidance given to him under this subsection.

(2) In sections 34A and 34B and this section—

"course organiser", in relation to a course, means the person who, in accordance with regulations made by the Secretary of State, is responsible for giving the certificates mentioned in section 34B(1) in respect of the completion of the course;

"petty sessions area" has the same meaning as in the Magistrates' Courts Act 1980;

"supervising court", in relation to an order under section 34A, means—

(a) in England and Wales, a magistrates' court acting for the petty sessions area named in the order as the area where the offender resides or will reside;

(b) in Scotland, the sheriff court for the district where the offender resides or will reside or, where the order is made by a stipendiary magistrate and the offender resides or will reside within his commission area, the district court for that area,

and any reference to the clerk of a magistrates' court is a reference to the clerk to the justices for the petty sessions area for which the court acts.

(3) Any power to make regulations under section 34B or this section—

(a) includes power to make different provision for different cases, and to make such incidental or supplemental provision as appears to the Secretary of State to be necessary or expedient;

(b) shall be exercisable by statutory instrument, which shall be subject to annulment in pursuance of a resolution of either House of Parliament."

31. Experimental period for section 30.

(1) Subject to the following provisions, no order shall be made under section 34A of the Road Traffic Offenders Act 1988 after the end of 1997 or such later time as may be specified in an order made by the Secretary of State.

(2) At any time before the restriction imposed by subsection (1) above has taken effect, the Secretary of State may by order provide that it shall not do so.

(3) In this section "the experimental period" means the period beginning when section 30 above comes into force and ending—

(a) when the restriction imposed by subsection (1) above takes effect, or

(b) if the Secretary of State makes an order under subsection (2) above, on a date specified in the order (being a date falling before the time when the restriction imposed by subsection (1) above would otherwise have taken effect).

(4) During the experimental period—

(a) no order shall be made under section 34A of the Road Traffic Offenders Act 1988 by virtue of a person's conviction under section 3A of the Road Traffic Act 1988, and

(b) no order shall be made under section 34A of the Road Traffic Offenders Act 1988 except by a magistrates' court acting for a petty sessions area (or in Scotland, a sheriff court for a district or a stipendiary magistrate for a commission area) which is for the time being designated for the purposes of this section.

(5) In relation to orders made under section 34A during the experimental period, that section shall have effect with the omission of subsection (6) and

section 34B shall have effect as if references to the supervising court were references to the court which made the order.

(6) The power to designate an area or district for the purposes of this section shall be exercisable by the Secretary of State by order, and includes power to revoke any designation previously made.

(7) An order under subsection (6) above shall specify the period for which an area or district is designated, and may—

(a) specify different periods for different areas or districts, and

(b) extend or abridge any period previously specified.

(8) The power to make an order under subsection (1) above shall not be exercisable after the end of 1997, and no more than one order may be made under that subsection.

(9) Any power of the Secretary of State to make orders under this section shall be exercisable by statutory instrument, and—

(a) no order shall be made under subsection (1) or (2) above unless a draft of it has been laid before and approved by resolution of each House of Parliament, and

(b) any statutory instrument containing an order under subsection (6) above shall be subject to annulment in pursuance of a resolution of either House.

32. Disqualification until test is passed.

For section 36 of the Road Traffic Offenders Act 1988 there shall be substituted—

36. "Disqualification until test is passed.

(1) Where this subsection applies to a person the court must order him to be disqualified until he passes the appropriate driving test.

(2) Subsection (1) above applies to a person who is disqualified under section 34 of this Act on conviction of—

(a) manslaughter, or in Scotland culpable homicide, by the driver of a motor vehicle, or

(b) an offence under section 1 (causing death by dangerous driving) or section 2 (dangerous driving) of the Road Traffic Act 1988.

(3) Subsection (1) above also applies—

(a) to a person who is disqualified under section 34 or 35 of this Act in such circumstances or for such period as the Secretary of State may by order prescribe, or

(b) to such other persons convicted of such offences involving obligatory endorsement as may be so prescribed.

(4) Where a person to whom subsection (1) above does not apply is convicted of an offence involving obligatory endorsement, the court may order him to be disqualified until he passes the appropriate driving test (whether or not he has previously passed any test).

(5) In this section—

"appropriate driving test" means—

(a) an extended driving test, where a person is convicted of an offence involving obligatory disqualification or is disqualified under section 35 of this Act,

(b) a test of competence to drive, other than an extended driving test, in any other case, "extended driving test" means a test of competence to drive prescribed for the purposes of this section, and

"test of competence to drive" means a test prescribed by virtue of section 89(3) of the Road Traffic Act 1988.

(6) In determining whether to make an order under subsection (4) above, the court shall have regard to the safety of road users.

(7) Where a person is disqualified until he passes the extended driving test—

(a) any earlier order under this section shall cease to have effect, and

(b) a court shall not make a further order under this section while he is so disqualified.

(8) Subject to subsection (9) below, a disqualification by virtue of an order under this section shall be deemed to have expired on production to the Secretary of State of evidence, in such form as may be prescribed by regulations under section 105 of the Road Traffic Act 1988, that the person disqualified has passed the test in question since the order was made.

(9) A disqualification shall be deemed to have expired only in relation to vehicles of such classes as may be prescribed in relation to the test passed by regulations under that section.

(10) Where there is issued to a person a licence on the counterpart of which are endorsed particulars of a disqualification under this section, there shall also be endorsed the particulars of any test of competence to drive that he has passed since the order of disqualification was made.

(11) For the purposes of an order under this section, a person shall be treated as having passed a test of competence to drive other than an extended driving test if he passes a corresponding test conducted—

(a) under the law of Northern Ireland, the Isle of Man, any of the Channel Islands, another member State, Gibraltar or a designated country or territory (as defined by section 89(11) of the Road Traffic Act 1988), or

(b) for the purposes of obtaining a British Forces licence (as defined by section 88(8) of that Act);

and accordingly subsections (8) to (10) above shall apply in relation to such a test as they apply in relation to a test prescribed by virtue of section 89(3) of that Act.

(12) This section is subject to section 48 of this Act.

(13) The power to make an order under subsection (3) above shall be exercisable by statutory instrument; and no such order shall be made unless a draft of it has been laid before and approved by resolution of each House of Parliament.

(14) The Secretary of State shall not make an order under subsection (3) above after the end of 2001 if he has not previously made such an order."

33. Short periods of disqualification.

In section 37 of the Road Traffic Offenders Act 1988 (effect of order of disqualification) after subsection (1) there shall be inserted—

"(1A) Where—

(a) the disqualification is for a fixed period shorter than 56 days in respect of an offence involving obligatory endorsement, or

(b) the order is made under section 26 of this Act,
subsection (1) above shall not prevent the licence from again having effect at the end of the period of disqualification."

34. Conditional offer of fixed penalty.
For sections 75 to 77 of the Road Traffic Offenders Act 1988 (which relate to Scotland only) there shall be substituted—

"Conditional offer of fixed penalty

75. Issue of conditional offer.

(1) Where in England and Wales—

(a) a constable has reason to believe that a fixed penalty offence has been committed, and

(b) no fixed penalty notice in respect of the offence has been given under section 54 of this Act or fixed to a vehicle under section 62 of this Act,
a notice under this section may be sent to the alleged offender by or on behalf of the chief officer of police.

(2) Where in Scotland a procurator fiscal receives a report that—

(a) an offence specified in Schedule 3 to this Act has been committed,

(b) an offence specified in Schedule 5 to this Act has been committed,

(c) an offence referred to in paragraph (a) or (b) above has been committed, being an offence of causing or permitting a vehicle to be used by another person in contravention of any provision made or any restriction or prohibition imposed by or under any enactment, or

(d) an offence of aiding, abetting, counselling, procuring or inciting the commission of an offence referred to in this subsection, has been committed,
he may send a notice under this section to the alleged offender.

(3) Where in Scotland, a constable—

(a) on any occasion has reason to believe that a person he finds is committing or has on that occasion committed a fixed penalty offence, he may hand to that person,

(b) in any case has reason to believe that a fixed penalty offence has been committed, he or another person authorised in that respect by the chief constable may send to the alleged offender,
a notice under this section.

(4) Subsections (2) and (3) above shall not apply where a fixed penalty notice has been fixed to a vehicle under section 62 of this Act.

(5) A notice under this section is referred to in this section and sections 76 and 77 as a "conditional offer".

(6) Where a person issues a conditional offer, he must notify the justices' clerk, or in Scotland clerk of court, specified in it of its issue and its terms; and that clerk is referred to in this section and sections 76 and 77 as "the fixed penalty clerk".

(7) A conditional offer must—

(a) give such particulars of the circumstances alleged to constitute the offence to which it relates as are necessary for giving reasonable information about the alleged offence,

(b) state the amount of the fixed penalty for that offence, and

(c) state that proceedings against the alleged offender cannot be commenced in respect of that offence until the end of the period of twenty-eight days following the date on which the conditional offer was issued or such longer period as may be specified in the conditional offer.

(8) A conditional offer must indicate that if the following conditions are fulfilled, that is—

(a) within the period of twenty-eight days following the date on which the offer was issued, or such longer period as may be specified in the offer, the alleged offender—

(i) makes payment of the fixed penalty to the fixed penalty clerk, and

(ii) where the offence to which the offer relates is an offence involving obligatory endorsement, at the same time delivers his licence and its counterpart to that clerk, and

(b) where his licence and its counterpart are so delivered, that clerk is satisfied on inspecting them that, if the alleged offender were convicted of the offence, he would not be liable to be disqualified under section 35 of this Act,

any liability to conviction of the offence shall be discharged.

(9) For the purposes of the condition set out in subsection (8)(b) above, it shall be assumed, in the case of an offence in relation to which a range of numbers is shown in the last column of Part I of Schedule 2 to this Act, that the number of penalty points to be attributed to the offence would be the lowest in the range.

(10) The Secretary of State may by order provide for offences to become or (as the case may be) to cease to be offences in respect of which a conditional offer may be sent under subsection (2)(b) above, and may make such modifications of the provisions of this Part of this Act as appear to him to be necessary for the purpose.

(11) An offence committed by aiding, abetting, counselling, procuring or inciting the commission of an offence which is an offence involving obligatory endorsement is itself an offence involving obligatory endorsement for the purposes of the application of this Part of this Act in Scotland.

(12) In relation to licences which came into force before 1st June 1990, the references in subsection (8) above to the counterpart of a licence shall be disregarded.

76. Effect of offer and payment of penalty.

(1) This section applies where a conditional offer has been sent to a person under section 75 of this Act.

(2) No proceedings shall be brought against any person for the offence to which the conditional offer relates until—

(a) in England and Wales, the chief officer of police, or

(b) in Scotland, the procurator fiscal or (where the conditional offer was issued under section 75(3) of this Act) the chief constable,

receives notice in accordance with subsection (4) or (5) below.

(3) Where the alleged offender makes payment of the fixed penalty in

accordance with the conditional offer, no proceedings shall be brought against him for the offence to which the offer relates.

(4) Where—

(a) the alleged offender tenders payment in accordance with the conditional offer and delivers his licence and its counterpart to the fixed penalty clerk, but

(b) it appears to the clerk, on inspecting the licence and counterpart, that the alleged offender would be liable to be disqualified under section 35 of this Act if he were convicted of the offence to which the conditional offer relates,

then subsection (3) above shall not apply and the clerk must return the licence and its counterpart to the alleged offender together with the payment and give notice that he has done so to the person referred to in subsection (2)(a) or (b) above.

(5) Where, on the expiry of the period of twenty-eight days following the date on which the conditional offer was made or such longer perid as may be specified in the offer, the conditions specified in the offer in accordance with section 75(8)(a) of this Act have not been fulfilled, the fixed penalty clerk must notify the person referred to in subsection (2)(a) or (b) above.

(6) In determining for the purposes of subsection (4)(b) above whether a person convicted of an offence would be liable to disqualification under section 35, it shall be assumed, in the case of an offence in relation to which a range of numbers is shown in the last column of Part I of Schedule 2 to this Act, that the number of penalty points to be attributed to the offence would be the lowest in the range.

(7) In any proceedings a certificate that by a date specified in the certificate payment of a fixed penalty was or was not received by the fixed penalty clerk shall, if the certificate purports to be signed by that clerk, be evidence, or in Scotland sufficient evidence, of the facts stated.

(8) In relation to licences which came into force before 1st June 1990, the references in subsection (4) above to the counterpart of a licence shall be disregarded.

(9) In Scotland, the Secretary of State may by regulations vary the provisions of subsection (2)(b) above.

77. Endorsement where penalty paid.

(1) Where—

(a) in pursuance of a conditional offer a person (referred to in this section as the "licence holder") makes payment of the fixed penalty to the fixed penalty clerk and delivers his licence and its counterpart to the clerk, and

(b) the clerk is not required by subsection (4) of section 76 of this Act to return the licence and its counterpart to him and did not, before the payment was tendered, notify the person referred to in section 76(2)(a) or (b) of this Act under subsection (5) of that section,

the clerk must forthwith endorse the relevant particulars on the counterpart of the licence and return it to the licence holder together with the licence.

(2) Where it appears to a fixed penalty clerk in Scotland that there is an

error in an endorsement made by virtue of this section on the counterpart of a licence he may amend the endorsement so as to correct the error; and the amended endorsement shall have effect and shall be treated for all purposes as if it had been correctly made on receipt of the fixed penalty.

(3) Subject to subsection (4) below, where a cheque tendered in payment is subsequently dishonoured—

(a) any endorsement made by a clerk under subsection (1) above remains effective, notwithstanding that the licence holder is still liable to prosecution in respect of the alleged offence to which the endorsement relates, and

(b) the clerk must, upon the expiry of the period specified in the conditional offer or, if the period has expired, forthwith notify the person referred to in section 76(2)(a) or (b) of this Act that no payment has been made.

(4) When proceedings are brought against a licence holder after a notice has been given in pursuance of subsection (3)(b) above, the court—

(a) must order the removal of the fixed penalty endorsement from the counterpart of the licence, and

(b) may, on finding the licence holder guilty, make any competent order of endorsement or disqualification and pass any competent sentence.

(5) The reference in subsection (1) above to the relevant particulars is to—

(a) particulars of the offence, including the date when it was committed, and

(b) the number of penalty points to be attributed to the offence.

(6) The fixed penalty clerk must send notice to the Secretary of State—

(a) of any endorsement under subsection (1) above and of the particulars endorsed,

(b) of any amendment under subsection (2) above, and

(c) of any order under subsection (4)(a) above.

(7) Where the counterpart of a person's licence is endorsed under this section he shall be treated for the purposes of sections 13(4), 28, 29 and 45 of this Act and of the Rehabilitation of Offenders Act 1974 as if—

(a) he had been convicted of the offence,

(b) the endorsement had been made in pursuance of an order made on his conviction by a court under section 44 of this Act, and

(c) the particulars of the offence endorsed by virtue of subsection (5)(a) above were particulars of his conviction of that offence.

(8) In relation to any endorsement of the counterpart of a person's licence under this section—

(a) the reference in section 45(4) of this Act to the order for endorsement, and

(b) the references in section 13(4) of this Act to any order made on a person's conviction,

are to be read as references to the endorsement itself.

(9) In relation to licences which came into force before 1st June 1990, the references in this section to the counterpart of a licence shall be disregarded or, as the case may require, construed as references to the licence itself."

Miscellaneous

35. Disabled persons' badges.

(1) Section 21 of the Chronically Sick and Disabled Persons Act 1970 (badges for display on motor vehicles used by disabled persons) shall be amended in accordance with subsections (2) to (5) below.

(2) For subsections (2) and (3) there shall be substituted—

"(2) A badge may be issued to a disabled person of any prescribed description resident in the area of the issuing authority for one or more vehicles driven by him or used by him as a passenger."

(3) In subsection (4), the words "and any badge" onwards shall be omitted.

(4) After subsection (4) there shall be inserted—

"(4A) A badge issued under this section may be displayed only in such circumstances and in such manner as may be prescribed.

(4B) A person who drives a motor vehicle on a road (within the meaning of the Road Traffic Act 1988) at a time when a badge of a form prescribed under this section is displayed on the vehicle is guilty of an offence unless the badge is issued under this section and displayed in accordance with regulations made under it.

(4C) A person guilty of an offence under subsection (4B) above shall be liable on summary conviction to a fine not exceeding level 3 on the standard scale."

(5) In subsection (5), the words "and in the case" onwards shall be omitted.

(6) In section 117 of the Road Traffic Regulation Act 1984 (wrongful use of disabled person's badge) for subsections (1) and (2) there shall be substituted—

"(1) A person who at any time acts in contravention of, or fails to comply with, any provision of an order under this Act relating to the parking of motor vehicles is also guilty of an offence under this section if at that time—

(a) there was displayed on the motor vehicle in question a badge of a form prescribed under section 21 of the Chronically Sick and Disabled Persons Act 1970, and

(b) he was using the vehicle in circumstances where a disabled person's concession would be available to a disabled person's vehicle,

but he shall not be guilty of an offence under this section if the badge was issued under that section and displayed in accordance with regulations made under it."

36. Forfeiture of vehicles.

In section 43 of the Powers of Criminal Courts Act 1973 (power to deprive offender of property used, or intended for use, for purposes of crime) after subsection (1A) there shall be inserted—

"(1B) Where a person commits an offence to which this subsection applies by—

(a) driving, attempting to drive, or being in charge of a vehicle, or

(b) failing to comply with a requirement made under section 7 of the Road Traffic Act 1988 (failure to provide specimen for analysis or laboratory test) in the course of an investigation into whether the offender had committed an offence while driving, attempting to drive or being in charge of vehicle, or

(c) failing, as the driver of a vehicle, to comply with subsection (2) or (3) of section 170 of the Road Traffic Act 1988 (duty to stop and give information or report accident),

the vehicle shall be regarded for the purposes of subsection (1)(a) above (and subsection (4)(b) below) as used for the purpose of committing the offence (and for the purpose of committing any offence of aiding, abetting, counselling or procuring the commission of the offence).

(1C) Subsection (1B) above applies to—

(a) an offence under the Road Traffic Act 1988 which is punishable with imprisonment,

(b) an offence of manslaughter, and

(c) an offence under section 35 of the Offences against the Person Act 1861 (wanton and furious driving)."

37. Forfeiture of vehicles: Scotland.

(1) In each of sections 223 and 436 of the Criminal Procedure (Scotland) Act 1975 (forfeiture of property) after subsection (1) there shall be inserted—

"(1A) Where a person commits an offence to which this subsection applies by—

(a) driving, attempting to drive, or being in charge of a vehicle, or

(b) failing to comply with a requirement made under section 7 of the Road Traffic Act 1988 (failure to provide specimen for analysis or laboratory test) in the course of an investigation into whether the offender had committed an offence while driving, attempting to drive or being in charge of a vehicle, or

(c) failing, as the driver of a vehicle, to comply with subsections (2) or (3) of section 170 of the Road Traffic Act 1988 (duty to stop and give information or report accident),

the vehicle shall be regarded for the purposes of subsection (1)(a) above as used for the purpose of committing the offence."

(2) In section 223 of that Act after subsection (1A) there shall be inserted—

"(1B) Subsection (1A) above applies to—

(a) an offence under the Road Traffic Act 1988 which is punishable with imprisonment,

(b) an offence of culpable homicide."

(3) In section 436 of that Act after subsection (1A) there shall be inserted—

"(1B) Subsection (1A) above applies to an offence under the Road Traffic Act 1988 which is punishable with imprisonment."

38. Disqualification where vehicle used for assault.

(1) Section 44 of the Powers of Criminal Courts Act 1973 (disqualification by Crown Court where vehicle used for purposes of crime) shall be amended as follows.

(2) After subsection (1) there shall be inserted—

"(1A) This section also applies where a person is convicted by or before any court of common assault or of any other offence involving an assault (including an offence of aiding, abetting, counselling or procuring, or inciting to the commission of, an offence)."

(3) In subsection (2) after the words "this section applies" there shall be inserted the words "by virtue of subsection (1) above".

(4) After subsection (2) there shall be inserted—

"(2A) If in a case to which this section applies by virtue of subsection (1A) above the court is satisfied that the assault was committed by driving a motor vehicle, the court may order the person convicted to be disqualified, for such period as the court thinks fit, for holding or obtaining such a licence."

39. Disqualification in Scotland where vehicle used to commit offence.
After each of sections 223 and 436 of the Criminal Procedure (Scotland) Act 1975 there shall be added sections numbered 223A and 436A in the following terms—

"Disqualification in Scotland where vehicle used to commit offence

(1) Where a person is convicted of an offence (other than one triable only summarily) and the court which passes sentence is satisfied that a motor vehicle was used for the purpose of committing, or facilitating the commission of that offence, the court may order him to be disqualified for such period as the court thinks fit from holding or obtaining a licence to drive a motor vehicle granted under Part III of the Road Traffic Act 1988.

(2) A court which makes an order under this section disqualifying a person from holding or obtaining a licence shall require him to produce any such licence held by him and its counterpart.

(3) Any reference in this section to facilitating the commission of an offence shall include a reference to the taking of any steps after it has been committed for the purpose of disposing of any property to which it relates or of avoiding apprehension or detection.

(4) In relation to licences which came into force before 1st June 1990, the reference in subsection (2) above to the counterpart of a licence shall be disregarded."

40. Power to install equipment for detection of traffic offences.

(1) In Part V of the Highways Act 1980 immediately before section 96 there shall be inserted—

95A. "Power to install equipment for detection of traffic offences.
A highway authority may install and maintain on or near a highway structures and equipment for the detection of traffic offences."

(2) In Part IV of the Roads (Scotland) Act 1984 after section 49 there shall be inserted—

"Equipment for detection of traffic offences

49A. Power to install equipment for detection of traffic offences.
A roads authority may install and maintain on or near a road structures and equipment for the detection of traffic offences."

41. Variation of charges at off-street parking places.
After section 35B of the Road Traffic Regulation Act 1984 there shall be inserted—

35C. "Variation of charges at off-street parking places.

(1) Where an order under section 35(1)(iii) of this Act makes provision as to the charges to be paid in connection with the use of off-street parking places, the authority making that order may vary those charges by notice given under this section.

(2) The variation of any such charges by notice is not to be taken to prejudice any power to vary those charges by order under section 35 of this Act.

(3) The Secretary of State may by regulations make provision as to the procedure to be followed by any local authority giving notice under this section.

(4) The regulations may, in particular, make provision with respect to—

(a) the publication, where an authority propose to give notice, of details of their proposal;

(b) the form and manner in which notice is to be given; and

(c) the publication of notices.

(5) In giving any notice under this section a local authority shall comply with the regulations."

42. Variation of charges at designated parking places.

After section 46 of the Road Traffic Regulation Act 1984 (which deals with charges at, and regulation of, parking places) there shall be inserted—

46A. "Variation of charges at designated parking places.

(1) Where, by virtue of section 46 of this Act, any charges have been prescribed by a designation order or by an order under that section, the authority making that order may vary those charges by notice given under this section.

(2) The variation of any such charges by notice is not to be taken to prejudice any power to vary those charges by order under section 46 of this Act.

(3) The Secretary of State may by regulations make provision as to the procedure to be followed by any local authority giving notice under this section.

(4) The regulations may, in particular, make provision with respect to—

(a) the publication, where an authority propose to give notice, of details of their proposal;

(b) the form and manner in which notice is to be given; and

(c) the publication of notices.

(5) In giving any notice under this section a local authority shall comply with the regulations."

43. Permitted and special parking areas outside London.

(1) Schedule 3 shall have effect for the purpose of making provision with respect to areas ouside London corresponding to that made with respect to London, and areas within London, under sections 63 to 79 of this Act.

(2) In this section "London" has the same meaning as it has in Part II of this Act.

44. Parking attendants.

(1) After section 63 of the Road Traffic Regulation Act 1984, there shall be inserted—

"Parking attendants

63A. Parking attendants.

(1) A local authority may provide for the supervision of parking places within their area by individuals to be known as parking attendants.

(2) Parking attendants shall also have such other functions in relation to stationary vehicles as may be conferred by or under any other enactment.

(3) A parking attendant shall be—

(a) an individual employed by the authority; or

(b) where the authority have made arrangements with any person for the purposes of this section, an individual employed by that person to act as a parking attendant.

(4) Parking attendants in Greater London shall wear such uniform as the Secretary of State may determine when exercising prescribed functions, and shall not exercise any of those functions when not in uniform.

(5) In this section "local authority" and "parking place" have the meanings given by section 32(4) of this Act."

(2) In section 35 of that Act (provisions as to use of parking places provided under section 32 or 33), subsection (9) shall be omitted.

45. Variable speed limits.

(1) Section 84 of the Road Traffic Regulation Act 1984 (speed limits on roads other than restricted roads), shall be amended as follows.

(2) For subsection (1) there shall be substituted—

"(1) An order made under this subsection as respects any road may prohibit—

(a) the driving of motor vehicles on that road at a speed exceeding that specified in the order,

(b) the driving of motor vehicles on that road at a speed exceeding that specified in the order during periods specified in the order, or

(c) the driving of motor vehicles on that road at a speed exceeding the speed for the time being indicated by traffic signs in accordance with the order.

(1A) An order made by virtue of subsection (1)(c) above may—

(a) make provision restricting the speeds that may be indicated by traffic signs or the periods during which the indications may be given, and

(b) provide for the indications to be given only in such circumstances as may be determined by or under the order;

but any such order must comply with regulations made under subsection (1B) below, except where the Secretary of State authorises otherwise in a particular case.

(1B) The Secretary of State may make regulations governing the provision which may be made by orders of local authorities under subsection (1)(c) above, and any such regulations may in particular—

(a) prescribe the circumstances in which speed limits may have effect by virtue of an order,

(b) prescribe the speed limits which may be specified in an order, and
(c) make transitional provision and different provision for different cases."

(3) In subsection (3) for the words "under subsection (1)" there shall be substituted the words "made by virtue of subsection (1)(a)".

(4) At the end there shall be added—

"(6) Any reference in a local Act to roads subject to a speed limit shall, unless the contrary intention appears, be treated as not including a reference to roads subject to a speed limit imposed only by virtue of subsection (1)(b) or (c) above."

46. Tramcars and trolley vehicles.

(1) After section 141 of the Road Traffic Regulation Act 1984 (tramcars and trolley vehicles) there shall be inserted—

141A. "Tramcars and trolley vehicles: regulations.

(1) The Secretary of State may by regulations provide that such of the provisions mentioned in subsection (2) below as are specified in the regulations shall not apply, or shall apply with modifications—

(a) to all tramcars or to tramcars of any specified class, or
(b) to all trolley vehicles or to trolley vehicles of any specified class.

(2) The provisions referred to in subsection (1) above are the provisions of sections 1 to 14, 18 and 81 to 89 of this Act.

(3) Regulations under this section—

(a) may make different provision for different cases,
(b) may include such transitional provisions as appear to the Secretary of State to be necessary or expedient, and
(c) may make such amendments to any special Act as appear to the Secretary of State to be necessary or expedient in consequence of the regulations or in consequence of the application to any tramcars or trolley vehicles of any of the provisions mentioned in subsection (2) above.

(4) In this section—

"special Act" means a local Act of Parliament passed before the commencement of this section which authorises or regulates the use of tramcars or trolley vehicles;
"tramcar" includes any carriage used on any road by virtue of an order under the Light Railways Act 1896; and
"trolley vehicle" means a mechanically propelled vehicle adapted for use on roads without rails under power transmitted to it from some external source (whether or not there is in addition a source of power on board the vehicle)."

(2) After section 193 of the Road Traffic Act 1988 (exemptions for tramcars, trolley vehicles etc) there shall be inserted—

193A. "Tramcars and trolley vehicles.

(1) The Secretary of State may by regulations provide that such of the provisions mentioned in subsection (2) below as are specified in the regulations shall not apply, or shall apply with modifications—

(a) to all tramcars or to tramcars of any specified class, or

(b) to all trolley vehicles or to trolley vehicles of any specified class.

(2) The provisions referred to in subsection (1) above are the provisions of—

(a) sections 12, 40A to 42, 47, 48, 66, 68 to 73, 75 to 79, 83, 87 to 109, 143 to 165, 168, 170, 171, 178, 190 and 191 of this Act, and

(b) sections 1, 2, 7, 8, 22, 25 to 29, 31, 32, 34 to 48, 96 and 97 of the Road Traffic Offenders Act 1988 (provisions requiring warning of prosecution etc and provisions connected with the licensing of drivers).

(3) Regulations under this section—

(a) may make different provision for different cases,

(b) may include such transitional provisions as appear to the Secretary of State to be necessary or expedient, and

(c) may make such amendments to any special Act as appear to the Secretary of State to be necessary or expedient in consequence of the regulations or in consequence of the application to any tramcars or trolley vehicles of any of the provisions mentioned in subsection (2) above.

(4) In this section "special Act" means a local Act of Parliament passed before the commencement of this section which authorises or regulates the use of tramcars or trolley vehicles."

47. Applications for licences to drive hackney carriages etc.

(1) Part II of the Local Government (Miscellaneous Provisions) Act 1976 (including that Part as it applies in any area at the commencement of this section) shall have effect with the insertion of the following subsection after subsection (1) of each of section 51 (licensing of drivers of private hire vehicles) and section 59 (qualifications for drivers of hackney carriages)—

"(1A) For the purpose of satisfying themselves as to whether an applicant is a fit and proper person to hold a driver's licence, a council may send to the chief officer of police for the police area in which the council is situated—

(a) a copy of that person's application, and

(b) a request for the chief officer's observations;

and the chief officer shall respond to the request."

(2) Where any local Act contains a provision requiring a district council to be satisfied as to the fitness of an applicant to hold a licence to drive a private hire vehicle or a hackney carriage, the council may send to the chief officer of police for the police area in which the council is situated—

(a) a copy of that person's application, and

(b) a request for the chief officer's observations;

and the chief officer shall respond to the request.

48. Minor and consequential amendments.

Schedule 4 to this Act, which makes minor amendments and amendments consequential on the preceding provisions of this Act, shall have effect.

49. Omission of enactments not brought into force.

Parts II, III and IV of Schedule 2 to the Road Traffic (Consequential Provisions) Act 1988 (re-enactment or amendment of certain enactments not brought into force) shall be omitted.

PART II

TRAFFIC IN LONDON

Priority routes

50. Designation of priority routes.

(1) The Secretary of State may by order ("a priority route order") designate any road in London as a priority route.

(2) The Secretary of State shall exercise his powers under subsection (1) above so as to provide for a network of priority routes in London ("the priority route network") with a view to improving the movement of traffic.

(3) Before making a priority route order, the Secretary of State shall consult—

(a) the London authority within whose area the proposed priority route is;

(b) the relevant Commissioner or, if appropriate, both Commissioners; and

(c) London Regional Transport.

(4) Where it appears to the Secretary of State that the designation of any road as a priority route is likely to affect a road within the area of—

(a) a London authority other than that consulted under subsection (3)(a) above; or

(b) a county council,

he shall also consult that other London authority, or that county council, before making the proposed priority route order.

51. The Secretary of State's traffic management guidance.

(1) The Secretary of State shall issue to the London authorities and the Director guidance ("the Secretary of State's traffic management guidance") with respect to the management of traffic in London, and in particular with respect to priority routes and the priority route network.

(2) Any such guidance may—

(a) include provision—

(i) setting out the Secretary of State's objectives in designating priority routes; and

(ii) with respect to the role of the Director; and

(b) be varied at any time by the Secretary of State.

(3) Before issuing or varying any such guidance, the Secretary of State shall consult—

(a) such associations of London authorities (if any) as he thinks appropriate;

(b) the two Commissioners;

(c) the Disabled Persons Transport Advisory Committee; and

(d) London Regional Transport.

(4) In preparing any such guidance, the Secretary of State shall have regard to the needs of people with a disability.

52. The Traffic Director for London.

(1) The Secretary of State shall appoint a person to be known as the Traffic Director for London (in this Act referred to as "the Director").

(2) Schedule 5 to this Act shall have effect with respect to the Director.

(3) In addition to the specific duties imposed on him by this or any other enactment, the Director shall have the general duty—

(a) of co-ordinating the introduction and maintenance of traffic management measures taken by highway authorities in relation to priority routes established under this Part of this Act; and

(b) of monitoring the operation of those measures.

(4) The Director shall keep under review the manner in which the London authorities exercise their functions under Part III of the New Roads and Street Works Act 1991 in relation to priority routes or roads which, in his opinion, are likely to affect traffic using any priority route.

(5) The Secretary of State shall set objectives which he expects the Director to meet in exercising his functions.

(6) The Secretary of State shall publish, in such manner as he considers appropriate, any objectives which he sets under subsection (5) above.

(7) The Director shall exercise his functions—

(a) so as to meet any such objectives, so far as it is reasonably practicable for him to do so; and

(b) in accordance with any directions which the Secretary of State may from time to time see fit to give him.

(8) Any objectives set for the Director under subsection (5) above and any directions given to him under subsection (7) above may be specific or general.

(9) The Secretary of State shall publish, in such manner as he considers appropriate, any directions which he gives to the Director under subsection (7) above.

53. The Director's network plan.

(1) As soon as is reasonably practicable after first receiving a copy of the Secretary of State's traffic management guidance, the Director shall prepare and submit to him, and to each of the London authorities, his plans for the design and operation of the priority route network ("the network plan").

(2) The Director may divide the network plan into such parts as he considers appropriate and prepare and submit those parts separately.

(3) In preparing the network plan, or any part of it, the Director shall have regard to the Secretary of State's traffic management guidance and to the needs of people with a disability.

(4) Before submitting the network plan, or any part of it, the Director shall consult—

(a) the Secretary of State;

(b) the relevant Commissioner or, if appropriate, both Commissioners;

(c) any London authority within whose area there is any road which, in the opinion of the Director, is likely to be affected;

(d) such county councils (if any) as he thinks appropriate;

(e) such associations of London authorities (if any) as he thinks appropriate; and

(f) London Regional Transport.

(5) The network plan shall, in particular, include provision with respect to—

(a) the Director's overall objectives for particular priority routes;

(b) the traffic management measures which he expects to see taken in relation to priority routes in general or particular priority routes;

(c) the Director's requirements as to the timetable for the phased introduction of the priority route network; and

(d) the operation and maintenance of traffic management measures taken in respect of priority routes.

(6) The Director may from time to time vary the network plan, but before doing so he shall consult the persons mentioned in subsection (4) above.

(7) In preparing any variation, the Director shall have regard to the Secretary of State's traffic management guidance and to the needs of people with a disability.

(8) After varying the network plan, the Director shall submit it to the Secretary of State and to each of the London authorities.

(9) The Director shall—

(a) keep the network plan under review; and

(b) have regard to the desirability of varying it, particularly in the light of any further guidance issued by the Secretary of State under section 51 of this Act.

Local plans and trunk road local plans

54. Duty of London authorities to prepare local plans.

(1) Each London authority shall, after first receiving a copy of—

(a) the Secretary of State's traffic management guidance; and

(b) the network plan,

prepare a statement ("the local plan") of their proposals with respect to the operation of those priority routes which are within their area and with respect to which they are the highway authority.

(2) A local plan shall be in such form as may be specified by the Director.

(3) Where the Director prepares and submits the network plan in parts, subsection (1) above applies separately with respect to each part of the network plan.

(4) A local plan shall be prepared in accordance with the timetable set out in the network plan by virtue of section 53(5)(c) of this Act.

(5) Where the Secretary of State asks a London authority to make provision in their local plan with respect to a trunk road within their area which is a priority route, that authority may make, or (as the case may be) vary, their local plan so that it also has effect in relation to that trunk road.

(6) In preparing their local plan, a London authority shall have regard to—

(a) the Secretary of State's traffic management guidance; and

(b) the network plan.

(7) A London authority's local plan shall, in particular—

(a) indicate which of their powers under the Highways Act 1980 or the Road Traffic Regulation Act 1984 they propose to exercise in relation to the priority routes to which their plan relates and the manner in which they propose to exercise them;

(b) identify any orders made under the Act of 1984 which are, in their opinion, inconsistent with their plan and indicate their proposals for varying or revoking them;

(c) indicate—

(i) which of their powers under the Act of 1980 or the Act of 1984 they propose to exercise in relation to those other roads in their area which are (or would otherwise be) likely to affect, or be affected by, traffic using any of the priority routes to which their plan relates; and

(ii) the manner in which they propose to exercise them;

(d) indicate how the proposals referred to in paragraphs (a),(b) and (c) relate, in particular, to the needs of people with a disability;

(e) specify—

(i) the period which they consider will be required to implement their plan, on the assumption that it is approved by the Director; and

(ii) a timetable ("the local plan timetable") for implementing the different elements of their plan;

(f) specify a programme of maintenance of those traffic management measures which are derived from the exercise, on or in relation to the priority routes to which their plan relates, of powers under the Acts of 1980 and 1984;

(g) specify the amount of the expenditure which, in the opinion of the authority, they will incur as a direct result of implementing their plan; and

(h) deal with any other matter which they consider relevant to the proper and effective implementation of their plan.

(8) In preparing their local plan, a London authority shall consult—

(a) the relevant Commissioner or, if appropriate, both Commissioners;

(b) London Regional Transport;

(c) such organisations representing the interests of people with a disability who may be affected by the plan as appear to the authority to be appropriate; and

(d) any other London authority within whose area there is situated any road which is not a priority route but which is, in the authority's opinion, likely to be affected by any of the priority routes to which their plan relates.

(9) A London authority shall submit their local plan to the Director for his approval.

(10) The Director shall not approve a local plan unless he is satisfied—

(a) that it is consistent with the Secretary of State's traffic management guidance and with the network plan;

(b) in the case of any provision which is inconsistent with the network plan or the Secretary of State's traffic management guidance, that that provision is nevertheless appropriate for inclusion in the local plan;

(c) with the costing of the authority's proposals; and

(d) with the local plan timetable.

(11) Every London authority shall—

(a) keep their local plan under review; and

(b) consider whether it needs to be varied, particularly in the light of—

(i) any further guidance issued by the Secretary of State under section 51 of this Act; and

(ii) any variation of the network plan made by the Director under section 53(6) of this Act.

55. The Director's trunk road local plans.

(1) Where any priority route, or part of a priority route, is a trunk road, the Secretary of State may give a direction to the Director requiring him to prepare a

statement of the Director's proposals with respect to the operation of that priority route or of such part of it as may be specified in the direction.

(2) Subsection (1) above does not apply in relation to any trunk road in relation to which provision has been made by a London authority (under section 54(5) of this Act) in their local plan.

(3) A statement prepared under subsection (1) above is referred to in this Part of this Act as a "trunk road local plan".

(4) The Director may from time to time vary any trunk road local plan.

(5) In preparing any trunk road local plan or variation, the Director shall have regard to the Secretary of State's traffic management guidance and the network plan and shall consult—

(a) the Secretary of State;

(b) the relevant Commissioner or, if appropriate, both Commissioners;

(c) any London authority within whose area is situated—

(i) any priority route to which the trunk road local plan will apply; or

(ii) any road which is not a priority route but which, in the opinion of the Director, is likely to be affected by any priority route to which the trunk road local plan will apply;

(d) such organisations representing the interests of people with a disability who may be affected by the plan as appear to him to be appropriate; and

(e) London Regional Transport.

(6) Any trunk road local plan shall—

(a) indicate which powers under the Highways Act 1980 or the Road Traffic Regulation Act 1984 the Director proposes should be exercised in relation to the priority routes to which the plan relates and the manner in which he proposes they should be exercised;

(b) identify any orders made under the Act of 1984 which are, in his opinion, inconsistent with the plan and indicate his proposals for their variation or revocation;

(c) indicate—

(i) which powers under the Act of 1980 or the Act of 1984 he proposes should be exercised in relation to those other roads within London which are (or would otherwise be) likely to affect, or be affected by, traffic using any of the priority routes to which the plan relates; and

(ii) the manner in which he proposes they should be exercised;

(d) indicate how the proposals referred to in paragraphs (a), (b) and (c) relate, in particular, to the needs of people with a disability;

(e) specify—

(i) the period which he considers will be required to implement the plan; and

(ii) a timetable for implementing the different elements of the plan;

(f) specify a programme of maintenance of those traffic management measures, which are derived from the exercise, on or in relation to the priority routes to which the plan relates, of powers under the Acts of 1980 and 1984; and

(g) deal with any other matter which the Director considers relevant to the proper and effective implementation of the plan.

(7) The Director shall, in relation to each of his trunk road local plans—

(a) keep the plan under review; and

(b) consider whether it needs to be varied, particularly in the light of—

(i) any further guidance issued by the Secretary of State under section 51 of this Act; and

(ii) any variation of the network plan which he makes under section 53(6) of this Act.

56. The Minister's trunk road local plans.

(1) Where any priority route, or part of a priority route, is a trunk road with respect to which—

(a) no provision has been made in a local plan; and

(b) no direction has been given under section 55(1) of this Act,

the Secretary of State shall prepare a statement of his own proposals ("the Minister's trunk road local plan") with respect to the operation of that priority route or any part of it.

(2) A minister's trunk road local plan may be varied at any time by the Secretary of State.

(3) In preparing any such plan or variation, the Secretary of State shall consult—

(a) the Director;

(b) any London authority within whose area is situated—

(i) any priority route to which the plan will apply; or

(ii) any road which is not a priority route but which, in the opinion of the Secretary of State, is likely to be affected by any priority route to which the plan will apply;

(c) the relevant Commissioner or, if appropriate, both Commissioners;

(d) such organisations representing the interests of people with a disability who may be affected by the plan as appear to him to be appropriate; and

(e) London Regional Transport.

(4) A Minister's trunk road local plan shall, in particular—

(a) indicate which powers under the Highways Act 1980 or the Road Traffic Regulation Act 1984 the Secretary of State proposes should be exercised in relation to the priority routes to which the plan relates and the manner in which he proposes they should be exercised;

(b) identify any orders made under the Act of 1984 which are, in his opinion, inconsistent with the plan and indicate his proposals for their variation or revocation;

(c) indicate—

(i) which powers under the Act of 1980 or the Act of 1984 he proposes should be exercised in relation to those other roads within London which are (or would otherwise be) likely to affect, or be affected by, traffic using any of the priority routes to which the plan relates; and

(ii) the manner in which he proposes they should be exercised;

(d) indicate how the proposals referred to in paragraphs (a), (b) and (c) relate, in particular, to the needs of people with a disability;

(e) specify—

(i) the period which he considers wll be required to implement the plan; and

(ii) a timetable for implementing the different elements of the plan;

(f) specify a programme of maintenance of those traffic management measures which are derived from the exercise, on or in relation to the priority routes to which the plan relates, of powers under the Acts of 1980 and 1984; and

(g) deal with any other matter which he considers relevant to the proper and effective implementation of the plan.

(5) Where the Secretary of State considers that the implementation of any part of the plan requires a London authority to exercise any of its powers he may, in writing, ask the authority to exercise such powers as he may specify in his request.

(6) Where—

(a) the Secretary of State has sent such a request to a London authority; but

(b) the authority have not, in his opinion, exercised the powers in question within a reasonable period,

the Secretary of State may direct them to do so.

(7) Where a London authority have failed to comply with a direction under subsection (6) above within such period as the Secretary of State considers could reasonably be required by them, he may himself exercise the powers in question.

(8) Anything done by the Secretary of State in the exercise of those powers shall be treated for all purposes as if it had been done by the authority.

(9) Where the Secretary of State proposes to exercise any of the powers of a London authority by virtue of subsection (7) above, he may direct that authority not to exercise those or any other such powers, in such circumstances or in relation to such matters, as may be specified in the direction.

(10) Where, having intervened under subsection (7) above, the Secretary of State is satisfied that continued intervention by him is unnecessary—

(a) he shall notify the authority accordingly in writing; and

(b) with effect from the date on which that notice is served by him, any direction given by him with respect to his intervention shall cease to have effect.

(11) Any reasonable administrative expenses incurred by the Secretary of State in the exercise of his powers under subsection (7) above shall be recoverable by him from the London authority as a civil debt.

57. Implementation of local plans.

(1) Where the Director has approved a London authority's local plan, or has himself prepared a local plan on behalf of a London authority under section 61 of this Act, it shall be the duty of that authority to—

(a) implement the plan as soon as is reasonably practicable; and

(b) continue to act in a manner which is compatible with it.

(2) Every London authority shall provide the Director with such information, in such form and manner, as he may reasonably require with respect to the implementation or otherwise of their local plan.

(3) Where a London authority's local plan has effect in relation to a trunk road, by virtue of section 54(5) of this Act, the duty imposed by subsection (1) above shall apply in relation to the plan so far as it has that effect only if the Director, with the consent of the Secretary of State, gives a direction to that effect.

58. Implementation by Director of certain plans.

(1) Where the Secretary of State gives a direction to the Director requiring him to implement any trunk road local plan, or Minister's trunk road local plan, or part of any such plan, it shall be the duty of the Director to implement the provisions of the plan or (as the case may be) of that part of the plan, so far as they have effect in relation to any trunk road, as soon as is reasonably practicable.

(2) Any direction given under subsection (1) above may require any provision to which it applies to be implemented to such limited extent as may be specified in the direction.

(3) In so doing, the Director shall have all the powers which the Secretary of State would have in relation to any trunk road with respect to which the plan has effect, so far as may be necessary or expedient for the purpose of implementing the provisions of the plan.

(4) Anything done by the Director in purported exercise of those powers shall be taken to have been done by the Secretary of State.

(5) Where the Director considers that the implementation of any part of the plan requires a London authority to exercise any of its powers he may, in writing, ask the authority to exercise such powers as he may specify in his request.

(6) Where—

(a) the Director has sent such a request to a London authority; but

(b) the authority have not, in his opinion, exercised the powers in question within a reasonable period,

the Director may direct them to do so.

(7) Where a London authority have failed to comply with a direction under subsection (6) above within such period as the Director considers could reasonably be required by them, he may himself exercise the powers in question.

(8) Anything done by the Director in the exercise of those powers shall be treated for all purposes as if it had been done by the London authority.

(9) Where the Director proposes to exercise any of the powers of a London authority by virtue of subsection (7) above, he may direct that authority not to exercise those or any other such powers, in such circumstances or in relation to such matters, as may be specified in the direction.

(10) Where, having intervened under subsection (7) above, the Director is satisfied that continued intervention by him is unnecessary—

(a) he shall notify the London authority accordingly in writing; and

(b) with effect from the date on which that notice is served by him, any direction given by him with respect to his intervention shall cease to have effect.

(11) Any reasonable administrative expenses incurred by the Director in the exercise of his powers under subsection (7) above shall be recoverable by him from the London authority as a civil debt.

(12) Where the Secretary of State implements any of the provisions of a trunk road local plan, he shall have in relation to those provisions the powers conferred upon the Director by subsections (5) to (11) above.

59. Variation of local plans.

(1) A London authority may vary their local plan, but only with the written consent of the Director.

(2) The Director may give a direction to any London authority requiring them to vary their local plan in such manner as may be specified in the direction.

(3) In varying their local plan, a London authority shall have regard to the Secretary of State's traffic management guidance and to the network plan.

(4) Before varying their local plan, a London authority shall consult—

(a) the relevant Commissioner or, if appropriate, both Commissioners;

(b) London Regional Transport;

(c) such organisations representing the interests of people with a disability who may be affected by the plan as appear to the authority to be appropriate; and

(d) any other London authority within whose area there is situated any road which is not a priority route but which is, in the authority's opinion, likely to be affected by the proposed variation.

(5) Where a London authority fail, within a reasonable time, to comply with any direction given under subsection (2) above, the Director may vary the local plan on their behalf.

(6) Before varying a local plan on behalf of a London authority the Director shall consult—

(a) that authority;

(b) the relevant Commissioner or, if appropriate, both Commissioners;

(c) London Regional Transport;

(d) such organisations representing the interests of people with a disability who may be affected by the plan as appear to the Director to be appropriate; and

(e) any other London authority within whose area there is situated any road which is not a priority route but which is, in his opinion, likely to be affected by the proposed variation.

(7) Any reasonable administrative expenses incurred by the Director under subsection (5) above shall be recoverable by him from the London authority concerned as a civil debt.

60. Proposed action by London authorities likely to affect priority routes.

(1) No London authority shall exercise any power under the Highways Act 1980 or the Road Traffic Regulation Act 1984, in a way which will affect, or be likely to affect, a priority route unless the requirements of subsection (3) below have been satisfied.

(2) Subsection (1) above does not apply where the exercise of the power—

(a) accords with the provisions of the authority's approved local plan; or

(b) is in response to a request made, or direction given, under this Act by the Director or the Secretary of State.

(3) The requirements mentioned in subsection (1) above are that—

(a) the authority have given notice to the Director, in such manner as he may require, of their proposal to exercise the power in the way in question; and

(b) either—

(i) the Director has approved their proposal; or

(ii) the period of one month beginning with the date on which he received notice of the proposal has expired without his having objected to it.

(4) The Secretary of State may by an instrument in writing exclude any power from the application of this section to the extent specified in the instrument.

(5) Any such instrument may, in particular, exclude a power as respects—

(a) all or any of the London authorities;

(b) all or any of the priority routes; or

(c) the exercise of the power in such manner or circumstances as may be specified in the instrument.

(6) If a London authority exercise any power in contravention of this section, the Director may take such steps as he considers appropriate to reverse or modify the effect of the exercise of that power.

(7) Any reasonable expenses incurred by the Director in taking any steps under subsection (6) above shall be recoverable by him from the London authority concerned as a civil debt.

61. Intervention powers.

(1) Where it appears to the Director that a London authority have failed—

(a) to prepare a local plan in accordance with the requirements of section 54 of this Act; or

(b) to submit their local plan to him in accordance with those requirements,

he may direct the authority to do so within such period as he may specify in the direction.

(2) Where the Director has given such a direction, but the London authority concerned have not complied with it within a reasonable time, he may himself prepare a local plan on their behalf.

(3) Where the Director refuses to approve a local plan under section 54 of this Act, the London authority concerned shall prepare and submit a new local plan under that section unless the Director serves written notice on them of his intention to exercise his powers under subsection (5) below.

(4) In preparing any local plan in compliance with subsection (3) above, the London authority shall comply with any directions given to them by the Director.

(5) If the Director—

(a) has refused to approve a local plan which has been prepared in accordance with the requirements of section 54 of this Act; and

(b) has served on the London authority concerned a notice of the kind mentioned in subsection (3) above,

he may himself prepare a local plan on behalf of that authority.

(6) Where the Director prepares a local plan on behalf of a London authority under this section—

(a) he shall consult—

(i) that authority;

(ii) the relevant Commissioner or, if appropriate, both Commissioners;

(iii) London Regional Transport;

(iv) such organisations representing the interests of people with a disability who may be affected by the plan as appear to the Director to be appropriate; and

(v) any other London authority within whose area there is situated any road which is not a priority route but which is, in his opinion, likely to be affected by any of the priority routes to which the plan relates; and

(b) any reasonable administrative expenses incurred by him in preparing the plan shall be recoverable by him from the authority as a civil debt.

62. Failure to implement local plans.

(1) Where it appears to the Director that a London authority—

(a) have not implemented, or are unlikely to implement, their local plan in accordance with the local plan timetable; or

(b) have not implemented, or are unlikely to implement, it in a satisfactory manner,

he may direct the authority to take such steps as are required to implement it in accordance with the local plan timetable, or (as the case may be) to implement it in a satisfactory manner, in accordance with such other timetable as he may draw up and specify in the direction.

(2) Where it appears to the Director that a London authority have acted in a manner which is incompatible with their local plan, he may direct them to take such steps as he considers appropriate with a view to securing, so far as is reasonably practicable, that the effects of that action are removed.

(3) Where a London authority have failed to comply with a direction under subsection (1) or (2) above, the Director may (with the consent of the Secretary of State) take any steps which still remain to be taken by the authority in accordance with the terms of the direction.

(4) The Secretary of State may limit his consent to the implementation by the Director of part only of the local plan, and where he does so the Director's powers under subsection (3) above shall be limited to implementing that part.

(5) For the purposes of enabling him to exercise the powers given to him by subsection (3) above, the Director shall have all the powers which the London authority concerned have in connection with the implementation of their local plan.

(6) Anything done by the Director in the exercise of those powers shall be treated for all purposes as if it had been done by the London authority.

(7) Where the Director proposes to exercise any of the powers of a London authority by virtue of subsection (5) above, he may direct that authority not to exercise those or any other powers, in such circumstances or in relation to such matters, as may be specified in the direction.

(8) Where, having intervened under subsection (3) above, the Director is satified that continued intervention by him is unnecessary—

(a) he shall notify the London authority accordingly in writing; and

(b) with effect from the date on which that notice is served by him, any direction given by him with respect to his intervention shall cease to have effect.

(9) Any reasonable administrative expenses incurred by the Director in the exercise of his powers under this section shall be recoverable by him from the London authority as a civil debt.

Parking in London

63. The Secretary of State's parking guidance.

(1) The Secretary of State shall issue guidance ("the Secretary of State's parking guidance") to the London authorities with a view to those authorities co-ordinating their action with respect to parking in London.

(2) It shall be the duty of the joint planning committee for London established under section 5 of the Local Government Act 1985—

(a) to make proposals to the Secretary of State (if it thinks fit) as to the content of the Secretary of State's parking guidance; and

(b) to keep that guidance under review, with a view to making from time to time such further proposals as it considers appropriate.

(3) Before issuing or varying any guidance under this section, the Secretary of State shall consult—

(a) the two Commissioners;

(b) London Regional Transport;

(c) the Disabled Persons Transport Advisory Committee;

(d) such associations of London authorities (if any) as he thinks appropriate; and

(e) such other persons (if any) as he thinks appropriate.

(4) In connection with the preparation of the Secretary of State's parking guidance regard shall be had to the needs of people with a disability.

(5) The Secretary of State's parking guidance may, in particular, include provision with respect to appropriate levels for—

(a) parking charges;

(b) penalty charges;

(c) charges made by London authorities for the removal, storage and disposal of vehicles; and

(d) charges in respect of the release of vehicles from immobilisation devices fixed under section 69 of this Act.

(6) The Secretary of State's parking guidance may be varied at any time by the Secretary of State.

64. Charges at designated parking places.

(1) In section 46 of the Road Traffic Regulation Act 1984 (charges at, and regulation of, designated parking places), in subsection (1) after the word "made" there shall be inserted the words "with respect to any parking place outside Greater London".

(2) After subsection (1) of that section there shall be inserted the following subsection—

"(1A) Subject to Parts I to III of Schedule 9 to this Act, where the authority by whom a designation order is made with respect to any parking place in Greater London impose charges to be paid for vehicles left in a parking place designated by the order, those charges shall be prescribed by the designation order or by a separate order made by the authority."

65. Contravention of certain orders relating to parking places in London not to be criminal offence.

(1) In section 47 of the Road Traffic Regulation Act 1984 (offences relating to designated parking places) the words "but this subsection does not apply in relation to any designated parking place in Greater London" shall be added at the end of subsection (1).

(2) In section 8 of that Act (contravention of orders under section 6 to be an offence), the following subsection shall be inserted after subsection (1)—

"(1A) Subsection (1) above does not apply in relation to any order under section 6 of this Act so far as it designates any parking places."

(3) The provisions of section 11 of that Act (contravention of experimental traffic order) shall become subsection (1) of that section and the following subsection shall be inserted as subsection (2)—

"(2) This section does not apply in relation to any experimental traffic order so far as it designates any parking places in Greater London."

66. Parking penalties in London.

(1) Where, in the case of a stationary vehicle in a designated parking place, a parking attendant has reason to believe that a penalty charge is payable with respect to the vehicle, he may—

(a) fix a penalty charge notice to the vehicle; or

(b) give such a notice to the person appearing to him to be in charge of the vehicle.

(2) For the purposes of this Part of this Act, a penalty charge is payable with respect to a vehicle, by the owner of the vehicle, if—

(a) the vehicle has been left—

(i) otherwise than as authorised by or under any order relating to the designated parking place; or

(ii) beyond the period of parking which has been paid for;

(b) no parking charge payable with respect to the vehicle has been paid; or

(c) there has, with respect to the vehicle, been a contravention of, or failure to comply with, any provision made by or under any order relating to the designated parking place.

(3) A penalty charge notice must state—

(a) the grounds on which the parking attendant believes that a penalty charge is payable with respect to the vehicle;

(b) the amount of the penalty charge which is payable;

(c) that the penalty charge must be paid before the end of the period of 28 days beginning with the date of the notice;

(d) that if the penalty charge is paid before the end of the period of 14 days beginning with the date of the notice, the amount of the penalty charge will be reduced by the specified proportion;

(e) that, if the penalty charge is not paid before the end of the 28 day period, a notice to owner may be served by the London authority on the person appearing to them to be the owner of the vehicle;

(f) the address to which payment of the penalty charge must be sent.

(4) In subsection (3)(d) above "specified proportion" means such proportion, applicable to all cases, as may be determined by the London authorities acting through the Joint Committee.

(5) A penalty charge notice fixed to a vehicle in accordance with this section shall not be removed or interfered with except by or under the authority of—

(a) the owner, or person in charge, of the vehicle; or

(b) the London authority for the place in which the vehicle in question was found.

(6) A person contravening subsection (5) above shall be guilty of an offence and liable on summary conviction to a fine not exceeding level 2 on the standard scale.

(7) Schedule 6 to this Act shall have effect with respect to penalty charges, notices to owners and other matters supplementing the provisions of this section.

67. Recovery of vehicles or of proceeds of disposal.

(1) Section 101 of the Road Traffic Regulation Act 1984 shall be amended as follows.

(2) In subsection (1) for "(5)" there shall be substituted "(5A)".

(3) In subsection (4) after the words "before a vehicle" there shall be inserted the words "found outside Greater London".

(4) After that subsection there shall be inserted—

"(4A) If, before a vehicle found in Greater London is disposed of by an authority in pursuance of subsections (1) to (3) above, the vehicle is claimed by a person who satisfies the authority that he is its owner and pays—

(a) any penalty charge payable in respect of the parking of the vehicle in the place from which it was removed; and

(b) such sums in respect of the removal and storage of the vehicle—

(i) as the authority may require; or

(ii) in the case of sums payable to a competent authority which is not a local authority, as may be prescribed,

the authority shall permit him to remove the vehicle from their custody within such period as they may specify or, where paragraph (b)(ii) applies, as may be prescribed."

(5) In subsection (5) after the words "which a vehicle" there shall be inserted the words "found outside Greater London".

(6) After that subsection there shall be inserted—

"(5A) If, before the end of the period of one year beginning with the date on which a vehicle found in Greater London is sold by an authority in pursuance of this section, any person satisfies that authority that at the time of the sale he was the owner of the vehicle, that authority shall pay him any sum by which the proceeds of sale exceed the aggregate of—

(a) any penalty charge payable in respect of the parking of the vehicle in the place from which it was removed; and

(b) such sums in respect of the removal, storage and disposal of the vehicle—

(i) as the authority may require; or

(ii) in the case of sums payable to a competent authority which is not a local authority, as may be prescribed."

(7) In subsection (6) for the words "and (5)" there shall be substituted the words "to (5A)".

68. Charges for removal, storage and disposal of vehicles.

(1) Section 102 of the Road Traffic Regulation Act 1984 shall be amended as follows.

(2) In subsection (2)—

(a) in paragraphs (b) and (c) after the words "local authority" there shall be inserted the words "other than a London authority"; and

(b) after paragraph (c) there shall be added—

"and

(d) a London authority shall be entitled to recover from any person responsible, such charges in respect of the removal, storage and disposal of a vehicle removed from a parking place designated under section 6, 9 or 45 of this Act or otherwise provided or controlled by that authority as they may require."

(3) In subsection (8)—

(a) in the definition of "appropriate authority", for paragraph (b) there shall be substituted—

"(b) in relation to a vehicle removed (by a person other than a constable or person acting in aid of a police force) from a place ouside Greater London, which is a parking place provided or controlled by a local authority, or from a place (not being a parking place) on a road or land in the open air, means the local authority in whose area that place is,";

(b) in that definition, the words following paragraph (b) shall be omitted; and

(c) at the end of that subsection there shall be added—

"and

"London authority" means any council of a London borough or the Common Council of the City of London."

(4) The following subsection shall be added at the end—

"(9) For the purposes of—

(a) subsection (2)(d) above, and

(b) paragraph (b) in the definition of "appropriate authority" in subsection (8) above,

a parking place provided under a letting or arrangements made by a local authority in pursuance of section 33(4) of this Act shall be treated as provided by that authority."

69. Immobilisation of vehicles in parking places.

(1) Where, in the case of a stationary vehicle in a designated parking place, a parking attendant has reason to believe that the vehicle has been permitted to remain at rest there in any of the circumstances specified in section 66(2) (a), (b) or (c) of this Act, he or another person acting under his direction may fix an immobilisation device to the vehicle.

(2) On any occasion when an immobilisation device is fixed to a vehicle in accordance with this section, the person fixing the device shall also fix to the vehicle a notice—

(a) indicating that such a device has been fixed to the vehicle and warning that no attempt should be made to drive it or otherwise put it in motion until it has been released from that device;

(b) specifying the steps to be taken in order to secure its release; and

(c) giving such other information as may be prescribed.

(3) A vehicle to which an immobilisation device has been fixed in accordance with this section may only be released from that device by or under the direction of a person authorised by the relevant authority to give such a direction.

(4) Subject to subsection (3) above, a vehicle to which an immobilisation device has been fixed in accordance with this section shall be released from that device on payment in any manner specified in the notice fixed to the vehicle under subsection (2) above of—

(a) the penalty charge payable in respect of the parking; and

(b) such charge in respect of the release as may be required by the relevant authority.

(5) A notice fixed to a vehicle in accordance with this section shall not be removed or interfered with except by or under the authority of—

(a) the owner, or person in charge, of the vehicle; or

(b) the relevant authority.

(6) A person contravening subsection (5) above shall be guilty of an offence and liable on summary conviction to a fine not exceeding level 2 on the standard scale.

(7) Any person who, without being authorised to do so in accordance with this section, removes or attempts to remove an immobilisation device fixed to a vehicle in accordance with this section shall be guilty of an offence and shall be liable on summary conviction to a fine not exceeding level 3 on the standard scale.

(8) In this section "relevant authority" means the London authority for the place in which the vehicle in question was found.

70. Exemptions from section 69.

(1) Section 69(1) of this Act shall not apply in relation to a vehicle if—

(a) a current disabled person's badge is displayed on the vehicle;

(b) not more than 15 minutes have elapsed since the end of any period for which the appropriate charge was duly paid at the time of parking; or

(c) not more than 15 minutes have elapsed since the end of any unexpired time (in respect of another vehicle) which is available at the relevant parking meter at the time of parking.

(2) In any case in which section 69(1) of this Act would apply to a vehicle but for subsection (1)(a) above and the vehicle was not, at the time at which it was parked, being used—

(a) in accordance with regulations under section 21 of the Chronically Sick and Disabled Persons Act 1970; and

(b) in circumstances falling within section 117(1)(b) of the Road Traffic Regulation Act 1984 (use where a disabled person's concession would be available),

the person in charge of the vehicle at that time shall be guilty of an offence and liable on summary conviction to a fine not exceeding level 3 on the standard scale.

(3) In this section "disabled person's badge" has the same meaning as in section 142(1) of the Road Traffic Regulation Act 1984, and "parking meter" has the same meaning as in section 46(2)(a) of that Act.

71. Representations in relation to removal or immobilisation of vehicles.

(1) The owner or person in charge of a vehicle who—

(a) removes it from the custody of a London authority in accordance with subsection (4A) of section 101 of the Road Traffic Regulation Act 1984 (ultimate disposal of vehicles abandoned and removable under that Act);

(b) receives any sum in respect of the vehicle under subsection (5A) of that section;

(c) is informed that the proceeds of sale of the vehicle did not exceed the aggregate amount mentioned in subsection (5A) of that section; or

(d) secures its release from an immobilisation device in accordance with section 69(4) of this Act, shall thereupon be informed of his right under this section to make representations to the relevant authority and of the effect of section 72 of this Act.

(2) The relevant authority shall give that information, or shall cause it to be given, in writing.

(3) Any person to whom subsection (1) above applies may make representations to the relevant authority on one or more of the grounds mentioned in subsection (4) below.

(4) The grounds are—

(a) that there were no reasonable grounds for the parking attendant concerned to believe that the vehicle had been permitted to remain at rest in the parking place by a person who was in control of the vehicle without the consent of the owner;

(c) that the place in which the vehicle was at rest was not a designated parking place;

(d) in a case within subsection (1)(d) above, that, by virtue of an exemption given by section 70 of this Act, section 69 of this Act did not apply to the vehicle at the time in question; or

(e) that the penalty or other charge in question exceeded the amount applicable in the circumstances of the case.

(5) An authority may disregard any representations which are received by them after the end of the period of 28 days beginning with the date on which the person making them is informed, under subsection (1) above, of his right to make representations.

(6) It shall be the duty of an authority to whom representations are duly made under this section, before the end of the period of 56 days beginning with the date on which they receive the representations—

(a) to consider them and any supporting evidence which the person making them provides; and

(b) to serve on that person notice of their decision as to whether they accept that the ground in question has been established.

(7) Where an authority serve notice under subsection (6)(b) above that they accept that a ground has been established they shall (when serving that notice) refund any sums—

(a) paid under subsection (4A) of section 101 of the Act of 1984 when the vehicle was removed from the custody of the authority;

(b) deducted from the proceeds of sale of the vehicle under subsection (5A) of that section; or

(c) paid under section 69(4) of this Act when the vehicle was released,

except to the extent (if any) to which those sums were properly paid or deducted.

(8) Where an authority serve notice under subsection (6)(b) above that they do not accept that a ground has been established, that notice shall—

(a) inform the person on whom it is served of his right to appeal to a parking adjudicator under section 72 of this Act;

(b) indicate the nature of a parking adjudicator's power to award costs against any person appealing to him under that section; and

(c) describe in general terms the form and manner in which such an appeal is required to be made.

(9) Where an authority fail to comply with subsection (6) above before the end of the period of 56 days mentioned there—

(a) they shall be deemed to have accepted that the ground in question has been established and to have served notice to that effect under subsection (7) above; and

(b) subsection (7) above shall have effect as if it required any refund to be made immediately after the end of that period.

(10) A person who makes any representation under this section or section 72 of this Act which is false in a material particular and does so recklessly or knowing it to be false in that particular is guilty of an offence.

(11) Any person convicted of an offence under subsection (10) above shall be liable on summary conviction to a fine not exceeding level 5 on the standard scale.

(12) Any notice required to be served under this section may be served by post.

(13) Where the person on whom any document is required to be served by subsection (6) above is a body corporate, the document is duly served if it is sent by post to the secretary or clerk of that body.

(14) In this section and in section 72 of this Act "relevant authority" has the same meaning as in section 69(8) of this Act.

72. Appeals to parking adjudicator in relation to decisions under section 71.

(1) Where an authority serve notice under subsection (6)(b) of section 71 of this Act that they do not accept that a ground on which representations were made under that section has been established, the person making those representations may, before—

(a) the end of the period of 28 days beginning with the date of service of that notice; or

(b) such longer period as a parking adjudicator may allow,

appeal to a parking adjudicator against the authority's decision.

(2) On an appeal under this section, the parking adjudicator shall consider the representations in question and any additional representations which are made by the appellant on any of the grounds mentioned in section 71(4) of this Act and, if he concludes—

(a) that any of the representations are justified; and

(b) that the relevant authority would have been under the duty imposed by section 71(7) of this Act to refund any sum if they had served notice that they accepted that the ground in question had been established,

he shall direct that authority to make the necessary refund.

(3) It shall be the duty of any authority to whom such a direction is given to comply with it forthwith.

73. Appointment of parking adjudicators by joint committee of the London authorities.

(1) The London authorities shall establish a single joint committee under section 101(5) of the Local Government Act 1972 ("the Joint Committee") before the end of the period of two months beginning with the date on which the Secretary of State first issues his guidance under section 63 of this Act.

(2) The functions conferred on the London authorities by this section and section 74 of this Act shall be discharged by the Joint Committee.

(3) The London authorities shall—

(a) with the consent of the Lord Chancellor, appoint persons to act as parking adjudicators for the purposes of this Part of this Act;

(b) provide accommodation and administrative staff for the parking adjudicators; and

(c) determine the places at which parking adjudicators are to sit.

(4) To be qualified for appointment as a parking adjudicator, a person must have a 5 year general qualification (within the meaning of section 71 of the Courts and Legal Services Act 1990).

(5) Each parking adjudicator shall be appointed for such term, not exceeding five years, as the London authorities may specify in relation to his appointment.

(6) On the expiry of his term of appointment, a parking adjudicator shall be eligible for re-appointment.

(7) A parking adjudicator may be removed from office only for misconduct or on the ground that he is unable or unfit to discharge his functions but shall otherwise hold and vacate office in accordance with the terms of his appointment.

(8) The expenses of the Joint Committee incurred in the discharge of functions conferred on the London authorities by this Act shall be defrayed by the London authorities in such proportions as they may decide or, in default of a decision by them, as may be determined by an arbitrator nominated by the Chartered Institute of Arbitrators on the application of the Joint Committee.

(9) The costs of any reference to arbitration under subsection (8) above shall be borne by the London authorities in equal shares.

(10) Where the Secretary of State is satisfied that there has been, or is likely to be, a failure on the part of the London authorities to agree on the proportions in which the expenses of the Joint Committee are to be defrayed by them under subsection (8) above he may give the Joint Committee such directions as he considers appropriate in order to require it to refer the matter to arbitration under that subsection.

(11) The Secretary of State shall by regulations make provision as to the procedure to be followed in relation to proceedings before parking adjudicators.

(12) The regulations may, in particular, include provision—

(a) as to the manner in which appeals to parking adjudicators are to be made or withdrawn;

(b) authorising an appeal to a parking adjudicator to be disposed of on the basis of written representations unless the appellant requests an oral hearing;

(c) prescribing the procedure to be followed before the hearing of an appeal by a parking adjudicator;

(d) requiring any such hearing to be held in public except in prescribed circumstances;

(e) as to the persons entitled to appear and be heard on behalf of the parties;

(f) requiring persons to attend to give evidence and to produce documents;

(g) as to evidence at the hearing;

(h) as to the adjournment of hearings;

(i) for the award of costs in prescribed circumstances;

(j) for the settlement of costs, by taxation (and in particular by taxation in a county court) or by some other prescribed method;

(k) authorising decisions of parking adjudicators to be reserved;
(l) authorising or requiring parking adjudicators—
(i) to revise or set aside decisions;
(ii) to revoke or vary orders made by them;
(m) requiring decisions of, and orders made by, parking adjudicators, to be recorded;
(n) as to the proof of decisions of, and orders made by, parking adjudicators;
(o) authorising the correction of clerical errors in records kept in accordance with the requirements of the regulations;
(p) requiring service of—
(i) notice of decisions of parking adjudicators;
(ii) copies of any orders made by such adjudicators; or
(iii) notice of any corrections made by parking adjudicators in their decisions or orders.

(13) Subject to any provision made by the regulations, a parking adjudicator may regulate his own procedure.

(14) If any person who is required to attend a hearing held by a parking adjudicator, or to produce any document to a parking adjudicator in accordance with any regulations under subsection (11) above, fails without reasonable excuse to do so, he shall be guilty of an offence and liable on summary conviction to a fine not exceeding level 2 on the standard scale.

(15) Any amount which is payable under an adjudication of a parking adjudicator shall, if a county court so orders, be recoverable by the person to whom the amount is payable, as if it were payable under a county court order.

(16) Subsection (15) above does not apply to a penalty charge which remains payable following an adjudication under paragraph 5 of Schedule 6 to this Act.

(17) In accordance with such requirements as may be imposed by the Joint Committee, each parking adjudicator shall make an annual report to the Joint Committee on the discharge of his functions.

(18) The Joint Committee shall make and publish an annual report in writing to the Secretary of State on the discharge by the parking adjudicators of their functions.

74. Fixing of certain parking and other charges for London.

(1) It shall be the duty of the London authorities to set the levels of additional parking charges to apply in London.

(2) Different levels may be set for different areas in London and for different cases or classes of case.

(3) In discharging their duties under this section the London authorities shall have regard to the Secretary of State's parking guidance.

(4) The London authorities shall submit to the Secretary of State, for his approval, the levels of additional parking charges which they propose to set under subsection (1) above.

(5) If—
(a) the London authorities fail to discharge their duty under subsection (1) above; or

(b) the Secretary of State does not approve the levels of additional parking charges proposed by the London authorities,
the levels of additional parking charges for London shall be set by regulations made by the Secretary of State.

(6) It shall be the duty of the London authorities to impose additional parking charges at the levels set in accordance with the provisions of this section.

(7) The London authorities shall publish, in such manner as the Secretary of State may determine, the levels of additional parking charges which they have set.

(8) In this section "additional parking charges" means penalty charges, charges made by London authorities for the removal, storage and disposal of vehicles and charges in respect of the release of vehicles from immobilisation devices fixed under section 69 of this Act.

75. Immobilisation of vehicles in London by police.
In the Road Traffic Regulation Act 1984, the following section shall be inserted after section 106—

106A. "Immobilisation of vehicles in London.

(1) Sections 104 and 105 of this Act shall extend throughout Greater London if the Secretary of State makes an order to that effect.

(2) If such an order is made, section 106 of this Act shall cease to apply in relation to Greater London when the order comes into force.

(3) Before such an order comes into force, section 106 of ths Act shall have effect as if in subsection (7) the words "or by the Traffic Director for London" were added at the end and as if the following subsection were inserted after subsection (7)—

"(7A) Before making an order under this section at the request of the Traffic Director for London, the Secretary of State shall consult the appropriate local authority."

(4) The power of the Secretary of State to make an order under this section shall be exercisable by statutory instrument which shall be subject to annulment in pursuance of a resolution of either House of Parliament."

76. Special parking areas.

(1) Where a London authority apply to the Secretary of State for an order to be made under this section, the Secretary of State may make an order designating the whole, or any part, of that authority's area as a special parking area.

(2) Before making an order under this section, the Secretary of State shall consult the relevant Commissioner or, if appropriate, both Commissioners.

(3) While an order under this section is in force, the following provisions shall cease to apply in relation to the special parking area designated by the order—

(a) section 8 of the Road Traffic Regulation Act 1984 (contravention of, or failure to comply with, an order under section 6 of that Act to be an offence), so far as it relates to the contravention of, or failure to comply with, any provision of such an order—

(i) prohibiting or restricting the waiting of vehicles on any road; or
(ii) relating to any of the matters mentioned in paragraph 7 or 8 of

Schedule 1 to that Act (conditions for loading or unloading, or delivery or collecting);

(b) section 11 of the Act of 1984 (contravention of, or failure to comply with, an experimental traffic order under section 9 of that Act to be an offence), so far as it relates to any contravention of, or failure to comply with, any provision of such an experimental traffic order—

(i) prohibiting or restricting the waiting of vehicles on any road; or

(ii) relating to any of the matters mentioned in paragraph 7 or 8 of Schedule 1 to that Act (conditions for loading or unloading, or delivery or collecting);

(c) section 16 of the Act of 1984 (contravention of a temporary restriction order or notice under section 14 of that Act to be an offence), so far as it relates to the contravention of any provision of an order or notice under section 14 of that Act which suspends any provision of an order made under section 45 or 46 of the Act of 1984;

(d) section 15 of the Greater London Council (General Powers) Act 1974 (parking of vehicles on verges, central reservations and footpaths etc. to be an offence);

(e) section 19 of the Road Trffic Act 1988 (parking of heavy vehicles on verges, central reservations and footpaths etc. to be an offence);

(f) section 21 of the Act of 1988 (prohibition of driving or parking on cycle tracks), so far as it makes it an offence to park a motor vehicle wholly or partly on a cycle track.

(4) The Secretary of State may by order amend subsection (3) above by adding further provisions (but only in so far as they apply in relation to stationary vehicles).

(5) Before making an order under subsection (4) above, the Secretary of State shall consult—

(a) the two Commissioners; and

(b) such associations of London authorities (if any) as he thinks appropriate.

77. Application of provisions in relation to special parking areas.

(1) This section applies in relation to any vehicle which is stationary in a special parking area (but which is not in a designated parking place) in circumstances in which an offence would have been committed with respect to the vehicle but for section 76(3) above.

(2) A penalty charge shall be payable with respect to the vehicle by the owner of the vehicle.

(3) Section 66 of, and Schedule 6 to, this Act shall apply in relation to penalty charges payable by virtue of subsection (2) above, but subject to such modifications (if any) as the Secretary of State considers it appropriate to make in the order designating the special parking area in question.

(4) Where a parking attendant has reason to believe that a penalty charge is payable with respect to the vehicle by virtue of subsection (2) above, he or another person acting under his direction may fix an immobilisation device to the vehicle.

(5) Subsections (2) to (8) of section 69 of this Act shall apply in relation to a device fixed to a vehicle under subsection (4) above, but subject to such

modifications (if any) as the Secretary of State considers it appropriate to make in the order designating the special parking area in question.

(6) An order under section 76 designating a special parking area may make such modifications of any provision of, or amended by, this Part of this Act as the Secretary of State considers appropriate in consequence of the provisions of section 76 or this section or of the order.

Miscellaneous

78. Enforcement.

(1) In this section—

"certificated bailiff", means any person authorised to act as such under subsection (6) below; and

"a Part II debt" means any sum which is—

(a) payable under, or by virtue of, any provision of this Part of this Act; and

(b) recoverable as if it were payable under a county court order.

(2) The Lord Chancellor may by order make provision—

(a) for warrants of execution in respect of Part II debts, or such class or classes of Part II debts as may be specified in the order, to be executed by certificated bailiffs;

(b) as to the requirements which must be satisfied before any person takes, with a view to enforcing the payment of—

(i) a Part II debt; or

(ii) such class or classes of Part II debts as may be so specified,

any other step of a kind specified by the order.

(3) Any such order may make such incidental and supplemental provision (including modifications of any enactment other than this Act) as the Lord Chancellor considers appropriate in consequence of the provision made by that order under subsection (2) above.

(4) The Lord Chancellor may by regulations make provision in connection with the certification of bailiffs under this section and the execution of warrants of execution by such bailiffs.

(5) The regulations may, in particular, make provision—

(a) as to the security (if any) to be required from certificated bailiffs;

(b) as to the fees and expenses payable with respect to executions by certificated bailiffs; and

(c) for the suspension or cancellation of certificates issued under this section and with respect to the effect of any such suspension or cancellation.

(6) For the purposes of this section, a person is a certificated bailiff if he is authorised to act as such by a certificate signed—

(a) by a judge assigned to a county court district; or

(b) in such circumstances as may be specified in regulations made by the Lord Chancellor, by a district judge.

(7) Any person who is not a certificated bailiff but who purports to levy a distress as such a bailiff, and any person authorising him to levy it, shall be deemed to have committed a trespass.

79. Application to Crown and visiting forces.

(1) Nothing in Part II of this Act applies in relation to any vehicle which—

(a) at the relevant time is used or appropriated for use for naval, military or airforce purposes;

(b) belongs to any visiting forces (within the meaning of the Visiting Forces Act 1952); or

(c) at the relevant time is used or appropriated for use, by any such forces.

(2) Sections 66 and 69 to 71 of this Act apply to—

(a) vehicles in the public service of the Crown which are required to be registered under the Vehicles (Excise) Act 1971 (other than those which are exempted by subsection (1)(a) above); and

(b) persons in the public service of the Crown.

80. Financial provisions.

(1) With a view to reimbursing (in whole or in part) reasonable costs incurred by any London authority under sections 54 to 59, 61 and 62 of this Act, the Director may make such payments to the authority as he considers appropriate.

(2) The Secretary of State may, with the consent of the Treasury, make such grants to the Director as he considers appropriate to enable the Director to discharge his functions.

81. Minor and consequential amendments.

The minor and consequential amendments set out in Schedule 7 to this Act shall have effect.

82. Interpretation of Part II.

(1) In this Part of this Act—

"Commissioner" means the Commissioner of Police of the Metropolis or the Commissioner of Police for the City of London;

"designated parking place" means a parking place in London which is designated as a parking place under an order made under section 6, 9 or 45 of the Road Traffic Regulation Act 1984;

"the Director" means the Traffic Director for London appointed under section 52 of this Act;

"immobilisation device" has the same meaning as in section 104(9) of the Road Traffic Regulation Act 1984;

"the Joint Committee" has the meaning given by section 73(1) of this Act;

"local plan" has the meaning given in section 54(1) of this Act;

"local plan timetable" has the meaning given in section 54(7)(e) of this Act;

"London" means the area comprising the areas of the London boroughs, the City of London and the Temples;

"London authority" means any council of a London borough or the Common Council of the City of London;

"Minister's trunk road local plan" has the meaning given in section 56(1);

"network plan" has the meaning given by section 53(1) of this Act;

"parking attendant" has the same meaning as in section 63A of the Road Traffic Regulation Act 1984 (which is inserted by section 44 of this Act);

"penalty charge" has the same meaning as in section 66 of this Act;

"prescribed" means prescribed by regulations made by the Secretary of State;
"priority route" means a road designated by a priority route order;
"priority route order" has the meaning given in section 50(1) of this Act;
"priority route network" has the meaning given in section 50(2) of this Act;
"road" has the same meaning as in the Road Traffic Regulation Act 1984;
"the Secretary of State's parking guidance" has the meaning given in section 63(1) of this Act;
"the Secretary of State's traffic management guidance" has the meaning given in section 51(1) of this Act;
"trunk road" has the same meaning as in section 10 of the Highways Act 1980;
"trunk road local plan" has the meaning given in section 55(3) of this Act;
"vehicle hiring agreement" and "vehicle-hire firm" have the same meanings as in section 66 of the Road Traffic Offenders Act 1988 (hired vehicles).

(2) For the purposes of this Part of this Act, the owner of a vehicle shall be taken to be the person by whom the vehicle is kept.

(3) In determining, for the purposes of this Part of this Act, who was the owner of a vehicle at any time, it shall be presumed that the owner was the person in whose name the vehicle was at that time registered under the Vehicles (Excise) Act 1971.

(4) Section 28 of the Chronically Sick and Disabled Persons Act 1970 (power to define "disability" and other expressions) shall apply in relation to this Part of this Act as it applies to that Act.

(5) In determining, for the purposes of any provision of this Part of this Act, whether a penalty charge has been paid before the end of a particular period, it shall be taken to be paid when it is received by the London authority concerned.

(6) Any power to make an order or regulations conferred by this Part shall be exercisable by statutory instrument.

(7) Any statutory instrument made under this Part of this Act shall be subject to annulment in pursuance of a resolution of either House of Parliament.

PART III
SUPPLEMENTARY

83. Repeals.

The enactments mentioned in Schedule 8 to this Act (which include enactments which are spent) are hereby repealed to the extent specified in the third column of that Schedule.

84. Commencement.

(1) The preceding sections of, and the Schedules to, this Act shall come into force on such day as the Secretary of State may appoint by order made by statutory instrument; and different days may be appointed for different purposes and in respect of different areas.

(2) An order under subsection (1) above may make such transitional provision as appears to the Secretary of State to be necessary or expedient.

85. Expenses

Any expenditure incurred by the Secretary of State under or by virtue of this Act shall be payable out of money provided by Parliament.

86. Extent.
Except in so far as it amends any enactment extending there, this Act does not extend to Northern Ireland.

87. Short title.
This Act may be cited as the Road Traffic Act 1991.

SCHEDULES

Section 22.

SCHEDULE 1
AMENDMENT OF SCHEDULE 1 TO THE ROAD TRAFFIC OFFENDERS ACT 1988

1. Schedule 1 to the Road Traffic Offenders Act 1988 (procedural requirements applicable in relation to certain offences) shall be amended as follows.
2. After paragraph 1 there shall be inserted—
"1A. Section 1 also applies to—
(a) an offence under section 16 of the Road Traffic Regulation Act 1984 consisting in the contravention of a restriction on the speed of vehicles imposed under section 14 of that Act,
(b) an offence under subsection (4) of section 17 of that Act consisting in the contravention of a restriction on the speed of vehicles imposed under that section, and
(c) an offence under section 88(7) or 89(1) of that Act (speeding offences)."
3. In paragraph 2, at the beginning of sub-paragraph (c) there shall be inserted the word "to".
4. The Table in that Schedule shall be amended as follows.
5. In the entries relating to sections 1 and 2 of the Road Traffic Act 1988 (reckless driving offences) in column 2, for the word "reckless" there shall be substituted the word "dangerous".
6. After the entry relating to section 3 of that Act there shall be inserted—

"RTA Section 3A	Causing death by careless driving when under influence of drink or drugs.	Section 11 of this Act."

7. In the entry relating to section 4 of that Act (driving a motor vehicle when unfit through drink or drugs etc) in column 2, for the words "motor vehicle" there shall be substituted the words "mechanically propelled vehicle".
8. In the entry relating to section 28 of that Act (reckless cycling) in column 2, for the word "reckless" there shall be substituted the word "dangerous".
9. After the entry relating to section 36 of that Act there shall be inserted—

"RTA section 40A	Using vehicle in dangerous condition etc.	Sections 11 and 12(1) of this Act.
RTA section 41A	Breach of requirement as to brakes, steering-gear or tyres.	Sections 11 and 12(1) of this Act.
RTA section 41B	Breach of requirement as to weight: goods and passenger	Sections 11 and 12(1) of this

vehicles. Act."

10. In the entry relating to section 42 of that Act, for the words in column 2 there shall be substituted the words "Breach of other construction and use requirements".

11. In the entry relating to section 71 of that Act (driving goods vehicle in contravention of prohibition etc) in column 2, the word "goods" in each place where it occurs shall be omitted.

12. In the entries relating to sections 87(1) and 87(2) of that Act (driving without a licence etc) in column 2, for the word "without" there shall be substituted the words "otherwise than in accordance with".

13. After the entry relating to section 87(2) of that Act there shall be inserted—

"RTA section 92(10)	Driving after making false declaration as to physical fitness.	Sections 6, 11 and 12(1) of this Act."

14. In the entry relating to section 94 of that Act (failure to notify Secretary of State about disability etc) in column 1, for the words "Section 94" there shall be substituted the words "Section 94(3)".

15. After that entry there shall be inserted—

"RTA section 94(3A)	Driving after such a failure.	Sections 6, 11 and 12(1) of this Act.
RTA section 94A	Driving after refusal of licence under section 92(3) or revocation under section 93.	Sections 6, 11 and 12(1) of this Act."

16. In the entry relating to section 164(6) of that Act (failing to produce driving licence to constable etc) in column 2 for the words "to constable" there shall be substituted the word "etc".

17. In the entry relating to section 174(1) or (6) of that Act (false statements etc), in column 1, for "(6)" there shall be substituted "(5)".

Section 26.

SCHEDULE 2

AMENDMENT OF SCHEDULE 2 TO THE ROAD TRAFFIC OFFENDERS ACT 1988

1. Part I of Schedule 2 to the Road Traffic Offenders Act 1988 (prosecution and punishment of offences) shall be amended as follows.

2. In the entry relating to section 16(1) of the Road Traffic Regulation Act 1984 (contravention of temporary prohibition or restriction) in columns 5 to 7 there shall be inserted—

"Discretionary if committed in respect of a speed restriction.	Obligatory if committed in respect of a speed restriction.	3-6 or 3 (fixed penalty)"

3. In the entry relating to section 17(4) of that Act (use of special road contrary to scheme or regulations), in column 7, for "3" there shall be substituted "3-6 or 3 (fixed penalty) if committed in respect of a speed restriction, 3 in any other case."

4. In the entry relating to section 89(1) of that Act (exceeding speed limit) in column 7, for "3" there shall be substituted "3-6 or 3 (fixed penalty)".

5. In the entry relating to section 1 of the Road Traffic Act 1988 (causing death by reckless driving)—

(a) in column 2 for the word "reckless" there shall be substituted the word "dangerous", and

(b) in column 7 for "4" there shall be substituted "3-11".

6. In the entry relating to section 2 of that Act (reckless driving)—

(a) in column 2, for the word "Reckless" there shall be substituted the word "Dangerous";

(b) for the words in column 5 there shall be substituted the word "Obligatory"; and

(c) for the words in column 7 there shall be substituted "3-11".

7. After the entry relating to section 3 of that Act there shall be inserted—

"RTA section 3A	Causing death by careless driving when under influence of drink or drugs.	On indictment.	5 years or a fine or both.	Obligatory.	Obligatory.	3-11".

8. In the entry relating to section 4(1) of that Act (driving or attempting to drive when unfit through drink or drugs) in column 7 for "4" there shall be substituted "3-11".

9. In the entry relating to section 4(2) of that Act (being in charge of a motor vehicle when unfit to drive) in column 2, for the words "motor vehicle" there shall be substituted the words "mechanically propelled vehicle".

10. In the entry relating to section 5(1)(a) of that Act (driving or attempting to drive with excess alcohol in breath, blood or urine) in column 7 for "4" there shall be substituted "3-11".

11. In the entry relating to section 7 of that Act (failing to provide specimen for analysis or laboratory test) in column 7 for the words "4 in case" there shall be substituted the words "3-11 in case".

12. In the entry relating to section 12 of that Act (motor racing and speed trials on public ways) in column 7 for "4" there shall be substituted "3-11".

13. After the entry relating to section 22 of that Act there shall be inserted—

"RTA section 22A.	Causing danger to road-users.	(a) Summarily. (b) On indictment.	(a) 6 months or the statutory maximum or both. (b) 7 years or a fine or both.	—	—	—

14. In the entry relating to section 23 of that Act (carrying passenger on motor-cycle contrary to that section), in column 7, for "1" there shall be substituted "3".

15. In the entry relating to section 28 of that Act (dangerous cycling)—

(a) in column 2 for the word "reckless" there shall be substituted the word "dangerous"; and

(b) in column 4, for the words "Level 3" there shall be substituted the words "Level 4".

16. In the entry relating to section 29 of that Act (careless and inconsiderate cycling), in column 4, for the words "Level 1" there shall be substituted the words "Level 3".

17. For the entry relating to section 42 of that Act (contravention of construction and use regulations) there shall be substituted—

"RTA Section 40A	Using vehicle in dangerous condition etc.	Summarily.	(a) Level 5 on the standard scale if committed in respect of a goods vehicle or a vehicle adapted to carry more than eight passengers. (b) Level 4 on the standard scale in any other case.	Discretionary.	Obligatory.	3
RTA section 41A	Breach of requirement as to brakes, steering-gear or tyres.	Summarily.	(a) Level 5 on the standard scale if committed in respect of a goods vehicle or a vehicle adapted to carry more than eight passengers. (b) Level 4 on the standard scale in any other case.	Discretionary.	Obligatory.	3
RTA section 41B	Breach of requirement as to weight: goods and passenger vehicles.	Summarily.	Level 5 on the standard scale.	—	—	—
RTA section 42	Breach of other construction and use requirements.	Summarily.	(a) Level 4 on the standard scale if committed in respect of a goods vehicle or a vehicle adapted to carry more than eight passengers. (b) Level 3 on the standard scale in any other case."	—	—	—

18. In the entries relating to section 68 and 71 of that Act, in column 2, the word "goods" in each place where it occurs shall be omitted.

19. For the entry relating to section 87(1) of that Act (driving without a licence) there shall be substituted—

"RTA section 87(1)	Driving otherwise than in accordance with a licence.	Summarily.	Level 3 on the standard scale.	Discretionary in a case where the offender's driving would not have been in accordance with any licence that could have been granted to him.	Obligatory in the case mentioned in column 5.	3-6"

20. In the entry relating to section 87(2) of that Act (causing or permitting to drive without a licence), in column 2 for the word "without" there shall be substituted the words "otherwise than in accordance with".

21. After the entry relating to section 92(7C) of that Act there shall be inserted—

"RTA section 92(10)	Driving after making false declaration as to physical fitness.	Summarily.	Level 4 on the standard scale.	Discretionary.	Obligatory.	3-6"

22. In the entry relating to section 94 of that Act (failure to notify Secretary of State about disability etc) for the words "Section 94" there shall be substituted the words "Section 94(3)".

23. After that entry there shall be inserted—

"RTA section 94(3A)	Driving after such a failure.	Summarily.	Level 3 on the standard scale.	Discretionary.	Obligatory.	3-6
RTA section 94A	Driving after refusal of licence under section 92(3) or revocation under section 93.	Summarily.	6 months or level 5 on the standard scale or both.	Discretionary.	Obligatory.	3-6"

24. In the entry relating to section 96 of that Act (driving with uncorrected defective eyesight or refusing to submit to test of eyesight) in column 7 for "2" there shall be substituted "3".

25. In the entry relating to section 103(1)(b) of that Act (obtaining licence, or driving, while disqualified) for the words in column 7 there shall be substituted "6".

26. In the entry relating to section 143 of that Act (using vehicle while uninsured or unsecured against third-party risks) in column 4 for the words "Level 4" there shall be substituted the words "Level 5".

27. In the entry relating to section 164 of that Act (failing to produce driving licence to constable etc) in column 2 for the words "to constable" there shall be substituted the word "etc".

28. In the entry relating to section 165 of that Act (failing to give constable certain information or to produce documents) the word "constable" shall be omitted.

29. In the entry relating to section 170(4) of that Act (failing to stop after accident or give particulars or report accident)—

(a) for the words in column 4 there shall be substituted "Six months or level 5 on the standard scale or both"; and

(b) in column 7 for "8-10" there shall be substituted "5-10".

30. In the entry relating to section 172 of that Act (failure of person keeping vehicle and others to give police information as to identity of driver etc in the case of certain offences) the following shall be inserted in columns 5 to 7—

"Discretionary, if committed otherwise than by virtue of subsection (5) or (11).	Obligatory, if committed otherwise than by virtue of subsection (5) or (11).	3"

31. In the entry relating to section 178 of that Act (taking etc in Scotland a motor vehicle without authority), the entries in columns 6 and 7 shall be omitted.

32.—(1) Part II of Schedule 2 to the Road Traffic Offenders Act 1988 (disqualification and endorsement in relation to manslaughter, certain offences of theft etc) shall be amended as follows.

(2) In the entry relating to manslaughter or culpable homicide, in column 4 for "4" there shall be substituted "3-11".

(3) The entries in columns 3 and 4 relating to stealing or attempting to steal a motor vehicle or to section 12 or 25 of the Theft Act 1968 shall be omitted.

Section 43

SCHEDULE 3
PERMITTED AND SPECIAL PARKING AREAS OUTSIDE LONDON

Permitted parking areas

1.—(1) Where an application for an order under this sub-paragraph is made to the Secretary of State—

(a) with respect to the whole, or any part, of their area, by a county council in England and Wales;

(b) with respect to the whole of their area, by a metropolitan district council;

(c) with respect to the whole of their areas, by two or more metropolitan district councils acting jointly;

(d) with respect to the whole, or any part, of their area, by a regional or islands council in Scotland;

(e) with respect to the whole, or any part, of their area, by a district council in Wales acting with the consent of the county council; or

(f) with respect to the whole, or any part, of the Isles of Scilly, by the Council of the Isles of Scilly,

he may make an order designating the whole, or any part, of the area to which the application relates as a permitted parking area.

(2) Before making any such application, a county council in Wales shall consult the district councils whose areas lie wholly or partly within the area to which the application relates.

(3) Before making an order under sub-paragraph (1) above, the Secretary of State shall consult the appropriate chief officer of police.

(4) While an order under sub-paragraph (1) above is in force, the following provisions shall cease to apply in relation to the permitted parking area designated by the order—

(a) section 35A(1) of the Road Traffic Regulation Act 1984 (offences), so far as it relates to the contravention of, or non-compliance with, any provision of an order made under section 35 of that Act (use of parking places) in relation to parking places provided under section 32(1)(b) of that Act (power of local authorities to provide free parking places on roads); and

(b) section 47(1) of the Act of 1984 (offences) in so far as it applies in relation to any designated parking place.

(5) The Secretary of State may by order amend sub-paragraph (4) above by adding further provisions (but only in so far as they apply in relation to stationary vehicles).

(6) Before making an order under sub-paragraph (5) above, the Secretary of State shall consult—

(a) such representatives of chief officers of police; and

(b) such associations of local authorities (if any),

as he considers appropriate.

Special parking areas

2.—(1) Where an application for an order under this sub-paragraph is made to the Secretary of State—

(a) with respect to the whole, or any part, of their area, by a county council in England and Wales;

(b) with respect to the whole, or any part, of their area, by a metropolitan district council;

(c) with respect to the whole, or any part, of their area, by a regional or islands council in Scotland; or

(d) with respect to the whole, or any part, of the Isles of Scilly, by the Council of the Isles of Scilly,

he may make an order designating the whole, or any part, of the area to which the application relates as a special parking area.

(2) Before making any such application, a county council in Wales shall consult the district councils whose areas lie wholly or partly within the area to which the application relates.

(3) Before making an order under sub-paragraph (1) above, the Secretary of State shall consult the appropriate chief officer of police.

(4) While an order under sub-paragraph (1) above is in force, the following provisions shall cease to apply in relation to the special parking area designated by the order—

(a) section 5 of the Road Traffic Regulation Act 1984 (contravention of a traffic regulation order under section 1 of that Act to be an offence), so far as it relates to the contravention of any provision of such an order prohibiting or restricting the waiting, or the loading and unloading, of vehicles;

(b) section 11 of the Act of 1984 (contravention of, or failure to comply with, experimental traffic order under section 9 of that Act), so far as it relates to the contravention of, or failure to comply with, any provision of such an order prohibiting or restricting the waiting, or the loading and unloading, of vehicles;

(c) section 129(6) of the Roads (Scotland) Act 1984 (parking of a motor vehicle wholly or partly on a cycle track to be an offence);

(d) section 19 of the Road Traffic Act 1988 (parking of heavy vehicles on verges, central reservations and footpaths etc. to be an offence);

(e) section 21 of the Act of 1988 (prohibition of driving or parking on cycle tracks), so far as it makes it an offence to park a motor vehicle wholly or partly on a cycle track.

(5) The Secretary of State may by order amend sub-paragraph (4) above by adding further provisions (but only in so far as they apply in relation to stationary vehicles).

(6) Before making an order under sub-paragraph (5) above, the Secretary of State shall consult—

(a) such representatives of chief officers of police; and

(b) such associations of local authorities (if any);

as he considers appropriate.

Control of parking in permitted and special parking areas

3.—(1) This paragraph applies in relation to any vehicle which is stationary in a permitted parking area, or special parking area, in circumstances in which an offence would have been committed with respect to the vehicle but for paragraph 1 or (as the case may be) paragraph 2 above.

(2) A penalty charge shall be payable with respect to the vehicle, by the owner of the vehicle.

(3) An order under paragraph 1 or 2 above designating a permitted parking area, or special parking area, may—

(a) provide for such provisions of Part II of this Act as the Secretary of State considers appropriate to apply, with such modifications (if any) as he considers appropriate, in relation to the permitted or special parking area in question; and

(b) make such modifications of any enactment, including any provision of this Act, as the Secretary of State considers appropriate in consequence of the provisions of paragraph 1 or 2 above, this paragraph or the order.

Orders under this Schedule

4.—(1) Any power to make an order conferred by this Schedule shall be exercisable by statutory instrument.

(2) Any such statutory instrument shall be subject to annulment in pursuance of a resolution of either House of Parliament.

Section 48.

SCHEDULE 4
MINOR AND CONSEQUENTIAL AMENDMENTS

The Transport Act 1968 (c. 73)

1. In section 82(8) of the Transport Act 1968 (powers of entry and inspection), for the words "section 68 of the Road Traffic Act 1988" there shall be substituted the words "section 66A of the Road Traffic Act 1988".

2. In section 99(8) of that Act (inspection of records), for the words from "a certifying" to "1988" there shall be substituted the words "an examiner appointed under section 66A of the Road Traffic Act 1988".

The Chronically Sick and Disabled Persons Act 1970 (c. 44)

3. In section 20(1) of the Chronically Sick and Disabled Persons Act 1970, in paragraph (b) (certain invalid carriages to be treated as not being motor vehicles for the purposes of the Road Traffic Act 1988 etc)—

(a) after the words "Road Traffic Act 1988" there shall be inserted the words ", except section 22A of that Act (causing danger to road users by interfering with motor vehicles etc),", and

(b) at the end of the paragraph there shall be added the words "and sections 1 to 4, 163, 170 and 181 of the Road Traffic Act 1988 shall not apply to it".

The Vehicles (Excise) Act 1971 (c. 10)

4. In section 5 of the Vehicles (Excise) Act 1971 (exemptions from duty in connection with vehicle testing etc) in subsection (3)—

(a) in the definition of "authorised person", for the words from "person authorised" to "so authorised" there shall be substituted the words "person who is, or is acting on behalf of, an examiner or inspector entitled to carry out examinations for the purposes of that section", and for the words "goods vehicle examiner" there shall be substituted the words "vehicle examiner", and

(b) for the definition of "goods vehicle examiner" there shall be substituted—

" "vehicle examiner" means an examiner appointed under section 66A of the Road Traffic Act 1988."

5. In Schedule 4A to that Act (duty on vehicles used for carrying exceptional loads)—

(a) in paragraph 1, for the words "section 42 of the Road Traffic Act 1972" there shall be substituted the words "section 44 of the Road Traffic Act 1988", and

(b) in paragraph 4, for the words "section 40" there shall be substituted the words "section 41", and for the words "the Road Traffic Act 1972" in each place where they occur there shall be substituted the words "the Road Traffic Act 1988".

The Road Traffic (Foreign Vehicles) Act 1972 (c. 27)

6. In section 1 of the Road Traffic (Foreign Vehicles) Act 1972 (power to prohibit driving of foreign goods vehicle) in subsection (6)(a) for sub-paragraphs (i) to (iii) there shall be substituted the words "section 40A of the Road Traffic Act 1988 (using vehicle in dangerous condition etc) or regulations under section 41 of that Act (construction, weight, equipment etc of motor vehicles and trailers),".

7. In section 2(3B) of that Act (provisions supplementary to section 1) for "72(9)" there shall be substituted "72A".

8. In section 7(1) of the Act (interpretation)—

(a) in the definition of "examiner", for the words following "means" there shall be substituted the words "an examiner appointed under section 66A of the Road Traffic Act 1988, or a constable authorised to act for the purposes of this Act by or on behalf of a chief officer of police", and

(b) in the definition of "official testing station" for "72(8)" there shall be substituted "72A".

9. In Schedule 2 to that Act (provisions relating to vehicles and their drivers) after the entry relating to section 100 of the Transport Act 1968 there shall be inserted the following entry—

"Section 40A of the Road Traffic Act 1988.	To create offence of using motor vehicle or trailer in dangerous condition etc."

The International Road Haulage Permits Act 1975 (c. 46)

10. In section 1(9) of the International Road Haulage Permits Act 1975 in the definition of "examiner" for the words "section 68(1)" there shall be substituted the words "section 66A".

The Highways Act 1980 (c. 66)

11. In section 42 of the Highways Act 1980 (power of district councils to maintain certain highways) in subsection (2)(c)(ii) for the words "under section 84 of that Act imposing a special limit" there shall be substituted the words "made by virtue of section 84(1)(a) of that Act imposing a speed limit".

12. In each of sections 90A(1) and 90B(1) of that Act (construction of road humps) at the beginning of paragraph (b) there shall be inserted the words "(whether or not the highway is subject to such a limit)".

13. In section 90F(2) of that Act (interpretation) for the definition of "statutory" there shall be substituted—

"statutory speed limit" means a speed limit having effect by virtue of an enactment other than section 84(1)(b) or (c) of the Road Traffic Regulation Act 1984 (temporary and variable speed limits)."

The Public Passenger Vehicles Act 1981 (c. 14)

14. In section 6(1)(a) of the Public Passenger Vehicles Act 1981 (certificates of fitness etc), for the words "a certifying officer" there shall be substituted the words "an examiner appointed under section 66A of the Road Traffic Act 1988".

15. In section 10(2) of that Act (approval of type vehicle), for the words "the certifying officer" there shall be substituted the words "an examiner appointed under section 66A of the Road Traffic Act 1988".

16.—(1) Section 51 of that Act (appeals to Secretary of State) shall be amended as follows.

(2) In subsection (1) for the words "a certifying officer" there shall be substituted the words "an examiner".

(3) In subsection (4) for the words "the certifying officer" in each place where they occur there shall be substituted the words "the examiner".

The Criminal Justice Act 1982 (c. 48)

17. In Part II of Schedule 1 to the Criminal Justice Act 1982 (offences excluded from Secretary of State's power to make orders concerning the early release of prisoners)—

(a) in the entry relating to section 1 of the Road Traffic Act 1988, for the word "reckless" there shall be substituted the word "dangerous", and

(b) after that entry there shall be inserted—

"Section 3A (causing death by careless driving when under the influence of drink or drugs)."

The Transport Act 1982 (c. 49)

18—(1) Section 9 of the Transport Act 1982 (private sector vehicle testing: the testing and surveillance functions) shall be amended as follows.

(2) Under the cross-heading "Functions under the 1988 Act"—

(a) for the paragraph beginning "The power of entry", there shall be substituted—

"The power of entry, inspection and detention of vehicles under section 68, but only in relation to vehicles brought to the place of inspection in pursuance of a direction given by a vehicle examiner or a constable under subsection (3) of that section."

(b) in the following paragraph, after "69" there shall be inserted "69A,", and for the word "goods vehicles" there shall be substituted the word "vehicles".

(3) Under the cross-heading "Functions under the 1981 Act", the paragraph beginning "Any functions under section 9" shall be omitted.

19.—(1) Section 10 of that Act (private sector vehicle testing: supplementary) shall be amended as follows.

(2) For subsection (3) there shall be substituted—

"(3) The words "or an authorised inspector" shall be inserted—

(a) in sections 51(1)(b) and 61(2)(a) of the 1988 Act, after the words "a vehicle examiner", and

(b) in section 6(1)(a) and 10(2) of the 1981 Act, after the words "Act 1988";

and the words "or authorised inspector" shall be inserted after the word "examiner" wherever occurring in section 69 of the 1988 Act."

(3) In subsection (6) for "68(3)" there shall be substituted "68(1)", and for "(4)" there shall be substituted "(3)".

(4) In subsection (9), in subsection (2A) to be inserted in section 20 of the Public Passenger Vehicles Act 1981, for the words "public service vehicle examiner" there shall be substituted the words "an examiner appointed under section 66A of the Road Traffic Act 1988".

(5) In subsection (10) for the words from 'certifying officer" to "goods vehicle examiner" there shall be substituted the words "vehicle examiner".

20. For section 20 of that Act (substitution of new section for section 72 of the 1988 Act) there shall be substituted—

"Amendment of section 72 of 1988 Act.

20. In section 72 of the 1988 Act (removal of prohibitions) after the word "constable" in each place where it occurs in subsections (1), (5) and (7), there shall be inserted the words "or authorised inspector".

21. For section 21(4) of that Act (amendments with respect to appeals) there shall be substituted—

"(4) In section 51 of that Act (appeals to the Secretary of State)—

(a) in subsection (1), after the word "examiner" there shall be inserted the words "or an authorised inspector";

(b) after subsection (1) there shall be inserted the following subsections—

"(1A) A person aggrieved by the refusal of the prescribed testing authority to approve a vehicle as a type vehicle under section 10 of this Act or by the withdrawal by that authority under that section of such approval may appeal to the Secretary of State.

(1B) On any appeal under subsection (1A) above, the Secretary of State shall cause an examination of the vehicle concerned to be made by an officer of the Secretary of State appointed by him for the purpose and shall make such determination on the basis of the examination as he thinks fit."; and

(c) in subsection (4) after the word "examiner" in both places where it occurs there shall be inserted the words "prescribed testing authority or authorised inspector concerned."."

22.—(1) Section 24 of that Act (falsification of documents) shall be amended as follows.

(2) In subsection (2) for paragraph (a) there shall be substituted—

"(a) in subsection (4) after the words "of this Act" there shall be inserted the words "or an authorised inspector appointed under section 8 of the Transport Act 1982";".

(3) In subsection (4), in section 66A(2) to be inserted in the Public Passenger Vehicles Act 1981, for the words "a certifying officer, a public service vehicle examiner" there shall be substituted the words "an examiner appointed under section 66A of the Road Traffic Act 1988".

23. In section 26 of that Act (interpretation) for the definition of "goods vehicle examiner" there shall be substituted—

"vehicle examiner" means an examiner appointed under section 66A of the 1988 Act."

The Road Traffic Regulation Act 1984 (c. 27)

24. In section 9 of the Road Traffic Regulation Act 1984 (experimental traffic orders) in subsection (1)(b), as substituted by the New Roads and Street Works Act 1991, for the words "83(2) or 84" there shall be substituted the words "or 83(2) or by virtue of sections 84(1)(a)".

25. In section 17(2) of that Act (traffic regulation on special roads) at the end there shall be added—

"(d) include provisions having effect in such places, at such times, in such manner or in such circumstances as may for the time being be indicated by traffic signs in accordance with the regulations."

26. In section 44 of that Act (control of off-street parking outside Greater London) in subsection (5) for the words "under section 84" there shall be substituted the words "made by virtue of section 84(1)(a)".

27. In section 49 of that Act (supplementary provisions as to designation orders and designated parking places), after subsection (4) there shall be inserted the following subsection—

"(4A) A constable, or a person acting under the instructions (whether general or specific) of the chief officer of police, may suspend the use of a parking place designated under section 45 of this Act for not more than 7 days in order to prevent or mitigate congestion or obstruction of traffic, or danger to or from traffic, in consequence of extraordinary circumstances."

28. In section 51 of that Act (parking devices), in subsection (5) the words "being not less than 2 years" shall be omitted.

29. In section 65 of that Act (powers and duties of highway authorities as to placing of traffic signs) after subsection (1) there shall be inserted—

"(1A) The power to give general directions under subsection (1) above includes power to require equipment used in connection with traffic signs to be of a type approved in accordance with the directions."

30. In section 85 of that Act (traffic signs for indicating speed restrictions) in subsections (1) and (2)(a) the words "the prescribed" shall be omitted.

31.—(1) Section 96 of that Act (additional powers of traffic wardens) shall be amended as follows.

(2) In subsection (2) at the end of paragraph (b) there shall be inserted—

"(bb) in this Act—

(i) section 100(3) (which relates to the interim disposal of vehicles removed under section 99); and

(ii) sections 104 and 105 (which relate to the immobilisation of illegally parked vehicles);".

(3) At the end of that section there shall be added—

"(4) Where an order has been made pursuant to subsection (2)(bb)(i) above, in section 100(3) of this Act the words "chief officer of the police force to which the constable belongs" shall be deemed to include a reference to a chief officer of police under whose direction a traffic warden acts.

(5) Any order made under section 95(5) of this Act may make different provision for different cases or classes of case, or in respect of different areas."

32. At the end of section 99 of that Act (removal of vehicles illegally parked) there shall be inserted—

"(6) For the purposes of this section, the suspension under section 13A or 49 of this Act of the use of a parking place is a restriction imposed under this Act."

33. In section 103 of that Act (supplementary provision as to removal of vehicles), for subsection (3) there shall be substituted—

"(3) Regulations made under sections 99 to 102 of this Act may make different provision for different cases or classes of case or in respect of different areas."

34.—(1) Section 104 of that Act (immobilisation of vehicles illegally parked) shall be amended as follows.

(2) In subsection (3) for the word "constable" there shall be substituted the words "person authorised to give such a direction by the chief officer of police within whose area the vehicle in question was found".

(3) In subsection (12) there shall be added at the end "or classes of case or in respect of different areas".

35. At the end of section 104 of that Act (immobilisation of vehicles illegally parked) there shall be inserted—

"(12A) For the purposes of this section, the suspension under section 13A or 49 of this Act of the use of a parking place is a restriction imposed under this Act."

36.—(1) Section 105 of that Act (exemptions from section 104) shall be amended as follows.

(2) In subsection (6)(a), for the words from "either" to "use) of" there shall be substituted the words "in accordance with regulations under".

(3) In subsection (6)(b), for "117(2)(b)" there shall be substituted "117(1)(b)".

37. In paragraph 13 of Schedule 9 to that Act (consent of Secretary of State before local authority make certain orders), after sub-paragraph (1)(d)(ii) there shall be inserted—

"(iii) a provision imposing a prohibition by virtue of paragraph (b) or (c) of that subsection, or".

The Roads (Scotland) Act 1984 (c. 54)

38.—(1) In section 36 of the Roads (Scotland) Act 1984 (construction of road humps by roads authority) at the beginning of paragraph (b) there shall be inserted the words "whether or not the road is subject to such a limit)".

(2) In section 40 of that Act (interpretation of sections 36 to 39) at the end of the definition of "statutory" there shall be added the words "other than section 84(1)(b) or (c) of the Road Traffic Regulation Act 1984 (temporary and variable speed limits)".

The Police and Criminal Evidence Act 1984 (c. 60)

39. In Part II of Schedule 5 to the Police and Criminal Evidence Act 1984 (serious arrestable offences)—

(a) in the entry relating to section 1 of the Road Traffic Act 1988, for the word "reckless" there shall be substituted the word "dangerous", and

(b) after that entry there shall be inserted—

"Section 3A (causing death by careless driving when under the influence of drink or drugs)."

The Coroners Act 1988 (c. 13)

40. In section 16 of the Coroners Act 1988 (adjournment of inquest in certain cases) in subsection (1)(a)(ii) for the words from "section" to "driving)" there shall be substituted the words "section 1 or 3A of the Road Traffic Act 1988 (dangerous driving or careless driving when under the influence of drink or drugs)".

41. In section 17 of that Act (supplementary provisions) in subsections (1)(b) and (2)(b) for the words from "section" to "driving)" there shall be substituted the words "section 1 or 3A of the Road Traffic Act 1988 (dangerous driving or careless driving when under the influence of drink or drugs)".

The Road Traffic Act 1988 (c. 52)

42. In section 7 of the Road Traffic Act 1988 (provision of specimens for analysis)—

(a) in subsection (1) for the words "section 4" there shall be substituted the words "section 3A, 4", and

(b) in subsection (3)(c) for the words "section 4" there shall be substituted the words "section 3A or 4".

43. In section 10 of that Act (detention of persons affected by alcohol or a drug) in subsections (1) and (2) for the words "motor vehicle" in each place where they occur there shall be substituted the words "mechanically propelled vehicle".

44. In section 11(1) of that Act (interpretation), for "4" there shall be substituted "3A".

45. In section 12 of that Act (motor racing on public ways), in subsection (2) for the words "public highway" there shall be substituted the word "highway".

46. In section 13 of that Act (regulation of motoring events on public ways), in subsection (4) for the words "public highway" there shall be substituted the word "highway".

47. In section 14 of that Act (seat belts: adults) in subsection (2)(b)(i) for the word "addresses" there shall be substituted the word "addressees".

48. In section 22 of that Act (leaving vehicles in dangerous positions) for the words "be likely to cause danger" there shall be substituted the words "involve a danger of injury".

49. In section 31 of that Act (regulation of cycle racing on public ways) for subsection (6) there shall be substituted—

"(6) In this section "public way" means, in England and Wales, a highway, and in Scotland, a public road but does not include a footpath."

50.—(1) Section 41 of that Act (regulation of construction, weight, equipment and use of vehicles) shall be amended as follows.

(2) In subsection (2) at the end of paragraph (e) there shall be added the words "(by means of the fixing of plates or otherwise) and the circumstances in which they are to be marked,".

(3) In subsection (2) after paragraph (j) there shall be inserted—

"(jj) speed limiters,".

(4) After subsection (4) there shall be inserted—

"(4A) Regulations under this section with respect to speed limiters may include provision—

(a) as to the checking and sealing of speed limiters by persons authorised in accordance with the regulations and the making of charges by them,

(b) imposing or providing for the imposition of conditions to be complied with by authorised persons,

(c) as to the withdrawal of authorisations."

51. In section 44(1) of that Act (authorisation of use on roads of special vehicles not complying with regulations under section 41) for the words from "and nothing" to "prevent" there shall be substituted the words "and sections 40A to 42 of this Act shall not apply in relation to".

52.—(1) Section 45 of that Act (tests of satisfactory condition of vehicles) shall be amended as follows.

(2) In subsection (1), for the words "prescribed statutory requirements" onwards there shall be substituted the words "following requirements are complied with, namely—

(a) the prescribed statutory requirements relating to the construction and condition of motor vehicles or their accessories or equipment, and

(b) the requirement that the condition of motor vehicles should not be such that their use on a road would involve a danger of injury to any person."

(3) In subsection (3), for paragraph (b) there shall be substituted—

"(b) examiners appointed under section 66A of this Act".

53. In section 46(a) of that Act (regulations as to authorisation of examiners), after the words "of examiners" there shall be inserted the words "in accordance with subsection (3)(a) of that section".

54.—(1) Section 49 of that Act (tests of satisfactory condition of goods vehicles and determination of plated weights etc) shall be amended as follows.

(2) In subsection (1), for the words following paragraph (b) there shall be substituted the words "or

(c) for the purpose of ascertaining whether the condition of the vehicle is such that its use on a road would involve a danger of injury to any person, or for any of those purposes."

(3) In subsection (2)(b), after the word "requirements" there shall be inserted the words "and the requirement that the condition of the vehicle is not such that its use on a road would involve a danger of injury to any person".

(4) In subsection (4), in the definition of "goods vehicle test", after the word "requirements" there shall be inserted the words ", or the requirement that the condition of the vehicle is not such that its use on a road would involve a danger of injury to any person,".

55.—(1) Section 50 of that Act (appeals against determinations) shall be amended as follows.

(2) In subsection (1) for the words "an area" onwards there shall be substituted the words "the Secretary of State".

(3) Subsections (2) and (3) shall be omitted.

56.—(1) Section 73 of that Act (provisions supplementary to sections 69 to 72) shall be amended as follows.

(2) For subsection (1) there shall be substituted—

"(1) Where it appears to a person giving a notice under section 69(6) or 70(2) of this Act that the vehicle concerned is an authorised vehicle, he must as soon as practicable take steps to bring the contents of the notice to the attention of—

(a) the traffic commissioner by whom the operator's licence was granted for the vehicle, and

(b) the holder of the licence if he is not in charge of the vehicle at the time when the notice is given.

(1A) Where it appears to a person giving a notice under section 69(6) or 70(2) of this Act that the vehicle concerned is used under a PSV operator's licence, he must as soon as practicable take steps to bring the contents of the notice to the attention of—

(a) the traffic commissioner by whom the PSV operator's licence was granted for the vehicle, and

(b) the holder of the licence if he is not in charge of the vehicle at the time when the notice is given.

(1B) In a case not within subsection (1) or subsection (1A) above, a person giving a notice under section 69(6) or 70(2) of this Act must as soon as practicable take steps to bring the contents of the notice to the attention of the owner of the vehicle if he is not in charge of it at the time when the notice is given.

(1C) A person giving a notice to the owner of a vehicle under section 72(7) of this Act must as soon as practicable take steps to bring the contents of the notice to the attention of any other person—

(a) who was the person to whom the previous notice under section 69(6) or 70(2) was given and was then the owner of the vehicle, or

(b) to whose attention the contents of the previous notice were brought under this section."

(3) Subsection (2) shall be omitted.

(4) In subsection (4) at the end there shall be added the words ", and "PSV operator's licence" has the same meaning as in the Public Passenger Vehicles Act 1981".

57. In subsection (1)(a) of section 74 of that Act (operator's duty to inspect goods vehicles) after the word "whether" there shall be inserted the words "the following requirements are complied with, namely—

(i)";

and for the words "are complied with" there shall be substituted the words "and

(ii) the requirement that the condition of the vehicle is not such that its use on a road would involve a danger of injury to any person".

58.—(1) Section 76 of that Act (fitting and supply of defective or unsuitable vehicle parts) shall be amended as follows.

(2) In subsection (1), after the words "to the vehicle" there shall be inserted the words "involve a danger of injury to any person or".

(3) In subsection (2)(b)(ii), after the words "its use" there shall be inserted the words "on a road", and at the end there shall be added the words "and would not involve a danger of injury to any person."

(4) At the end of each of subsection (3), (5)(b)(ii) and (6)(a) there shall be added the words "or involve a danger of injury to any person".

59.—(1) Section 79 of that Act (provisions relating to weighing of motor vehicles) shall be amended as follows.

(2) In subsection (2)—

(a) for "68(1)" there shall be substituted "66A";

(b) for the words from "vehicles of" to "vehicles generally" there shall be substituted the words "goods vehicles, public service vehicles, and vehicles which are not public service vehicles but are adapted to carry more than eight passengers,".

(3) In subsection (3)—

(a) for the wods from "vehicles of" to "vehicles generally" there shall be substituted the words "such vehicles", and

(b) for the words "a certifying officer," there shall be substituted the word "an".

60. In section 84(2) of that Act (remuneration of examiners), for the words from "goods" to "73" there shall be substituted the words "examiners appointed under section 66A".

61. In section 85 of that Act (interpretation of Part II)—

(a) in the definition of "official testing station" for "72(8)" there shall be substituted "72A", and

(b) after the definition of "prescribed" there shall be inserted—

"public service vehicle" has the same meaning as in the Public Passenger Vehicles Act 1981,".

62. In section 86 of that Act (index to Part II), in the table, after the entry for "Prescribed" there shall be inserted—

"Public service vehicle Section 85"

and after the entry for "Type approval requirements" there shall be inserted—

"Vehicle examiner Section 66A".

63. In section 89 of that Act (tests of competence to drive) at the end of subsection (3)(a) there shall be inserted the words "and section 36 of the Road Traffic Offenders Act 1988 (disqualification),".

64. In section 115(3) of that Act (revocation or suspension of large goods vehicle or passenger-carrying vehicle driver's licences) for the words "subsection

(1)(a) above" there shall be substituted the words "this section or section 117 of this Act".

65.—(1) Section 117 of that Act (disqualification on revocation of large goods vehicle or passenger-carrying vehicle driver's licences) shall be amended as follows.

(2) In subsection (1), for the words "for the purposes of that paragraph" there shall be substituted the words "in pursuance of section 115(3)".

(3) After subsection (2) there shall be inserted—

"(2A) Regulations may make provision for the application of subsections (1) and (2) above, in such circumstances and with such modifications as may be prescribed, where a person's large goods vehicle or passenger-carrying vehicle driver's licence is treated as revoked by virtue of section 37(1) of the Road Traffic Offenders Act 1988 (effect of disqualification by order of a court)."

66. In section 152 of that Act (duties of insurers etc: exceptions) at the end of subsection (2) there shall be added the words "and, for the purposes of this section, "material" means of such a nature as to influence the judgment of a prudent insurer in determining whether he will take the risk and, if so, at what premium and on what conditions."

67. In section 163(1) of that Act (power of police to stop vehicles) for the words "motor vehicle" there shall be substituted the words "mechanically propelled vehicle".

68.—(1) Section 164 of that Act (power of constable to require production of driving licence etc) shall be amended as follows.

(2) In subsection (1), after the word "constable" wherever it occurs there shall be inserted the words "or vehicle examiner".

(3) In subsection (2), for the words "Such a person" there shall be substituted the words "A person required by a constable under subsection (1) above to produce his licence".

(4) In subsection (3), after the word "constable" there shall be inserted the words "or vehicle examiner".

(5) In subsection (5) for the words "section 27 of the Road Traffic Offenders Act 1988" there shall be substituted the words "section 26 or 27 of the Road Traffic Offenders Act 1988 or section 44 of the Powers of Criminal Courts Act 1973 or section 223A or 436A of the Criminal Procedure (Scotland) Act 1975."

(6) In subsection (6) for the words "and (8)" there shall be substituted the words "to (8A)".

(7) After subsection (8) there shall be inserted—

"(8A) Subsection (8) above shall apply in relation to a certificate of completion of a training course for motor cyclists as it applies in relation to a licence."

(8) At the end of subsection (11) there shall be added the words "and "vehicle examiner" means an examiner appointed under section 66A of this Act."

69.—(1) Section 165 of that Act (powers of constables to obtain names and addresses of drivers etc) shall be amended as follows.

(2) In subsection (1), after the word "constable" wherever it occurs there shall be inserted the words "or vehicle examiner".

(3) In subsection (5), after the word "constable" wherever it occurs there shall be inserted the words "or vehicle examiner".

(4) At the end of subsection (7) there shall be added the words "and "vehicle examiner" means an examiner appointed under section 66A of this Act."

70. In section 166 of that Act (powers of certain officers as respects goods vehicles etc) for the words from the beginning to the end of paragraph (d) there shall be substituted the words "A person authorised for the purpose by a traffic commissioner appointed under the Public Passenger Vehicles Act 1981,".

71. In section 168 of that Act (offence of failing to give name and address in relation to certain offences) in paragraph (a) for the words "motor vehicle" there shall be substituted the words "mechanically propelled vehicle".

72.—(1) Section 170 of that Act (duty of driver to stop, report accident and give information or documents) shall be amended as follows.

(2) In subsections (1) to (3) for the words "motor vehicle" in each place where they occur there shall be substituted the words "mechanically propelled vehicle".

(3) In subsection (5) for the words "the vehicle" there shall be substituted the words "a motor vehicle".

(4) In subsection (7) for the word "five" there shall be substituted the word "seven".

73.—(1) Subsection (2) of section 173 of that Act (offences of forgery etc) shall be amended as follows.

(2) After paragraph (c) there shall be inserted—

"(cc) any seal required by regulations made under section 41 of this Act with respect to speed limiters,".

(3) In paragraph (d) for the words from the beginning to "Part II of this Act)" there shall be substituted the words "any plate containing particulars required to be marked on a vehicle by regulations under section 41 of this Act".

(4) After paragraph (d) there shall be inserted—

"(dd) any document evidencing the appointment of an examiner under section 66A of this Act,".

(5) After paragraph (f) there shall be inserted—

"(ff) any certificate provided for by regulations under section 97(3A) of this Act relating to the completion of a training course for motor cyclists,".

(6) After paragraph (l) there shall be added—

"and

(m) a certificate of the kind referred to in section 34B(1) of the Road Traffic Offenders Act 1988."

74.—(1) Section 176 of that Act (power to seize documents etc) shall be amended as follows.

(2) In subsection (4), for the words from "a certifying" to "68(1)" there shall be substituted the words "an examiner appointed under section 66A".

(3) In subsection (5)(a), for the words "for the purposes of sections 68 to 72" there shall be substituted the words "under section 66A".

75. In section 177 of that Act (impersonation of, or of person employed by, authorised examiner) after the words "a person authorised" there shall be inserted the words "in accordance with regulations made under section 41 of this

Act with respect to the checking and sealing of speed limiters or a person authorised".

76. In section 181 of that Act (provisions as to accident inquiries) in subsections (1) and (2) for the words "motor vehicle" in each place where they occur there shall be substituted the words "mechanically propelled vehicle".

77. In section 183(3) of that Act (Crown application), for the words from "68" to "1981" there shall be substituted the words "66A of this Act".

78.—(1) Section 192 of that Act (interpretation) shall be amended as follows.

(2) In subsection (1), in the definition of "road"—

(a) after the word ""road"" there shall be inserted "(a)", and

(b) at the end there shall be inserted—

"and

(b) in relation to Scotland, means any road within the meaning of the Roads (Scotland) Act 1984 and any other way to which the public has access, and includes bridges over which a road passes,".

(3) In subsection (1), in the definition of "trolley vehicle" for the words "and moved by" there shall be substituted the word "under", and at the end there shall be added the words "(whether or not there is in addition a source of power on board the vehicle)".

(4) In subsection (2) the word ""road"" shall be omitted.

79. In Schedule 4 to that Act (provisions not applicable to tramcars)—

(a) in paragraph 1, for "127" there shall be substituted "34",

(b) in paragraph 2, for the words "Sections 2, 3, 4(1) and 181 of this Act do not apply" there shall be substituted the words "Section 181 of this Act does not apply",

(c) in paragraph 3, for "41", there shall be substituted "40A to", and

(d) after paragraph 3 there shall be inserted—

"3A. Sections 68 and 69 of this Act do not apply to tramcars."

The Road Traffic Offenders Act 1988 (c. 53)

80. In section 1 of the Road Traffic Offenders Act 1988 (requirement of warning etc of prosecution of certain offences), in subsection (1) for the words "where a person" to "convicted unless" there shall be substituted the words "a person shall not be convicted of an offence to which this section applies unless".

81. For subsection (4) to (6) of section 2 of that Act (requirement of warning of prosecution: supplementary) there shall be substituted—

"(4) Failure to comply with the requirement of section 1(1) of this Act in relation to an offence is not a bar to the conviction of a person of that offence by virtue of the provisions of—

(a) section 24 of this Act, or

(b) any of the enactments mentioned in section 24(6);

but a person is not to be convicted of an offence by virtue of any of those provisions if section 1 applies to the offence with which he was charged and the requirement of section 1(1) was not satisfied in relation to the offence charged."

82. In section 5 of that Act (exemption from Licensing Act offence) for the words "section 4" there shall be substituted the words "section 3A, 4".

83. In section 7 of that Act (duty of accused to provide licence) for the words "obligatory endorsement" there shall be substituted the words "obligatory or discretionary disqualification".

84. In section 11(1) of that Act (evidence by certificate as to driver, user or owner) for the words "motor vehicle" in each place where they occur there shall be substituted the words "mechanically propelled vehicle".

85. In section 12 of that Act (proof of identity of driver) after subsection (3) there shall be added—

"(4) In summary proceedings in Scotland for an offence to which section 20(2) of the Road Traffic Act 1988 applies, where—

(a) it is proved to the satisfaction of the court that a requirement under section 172(2) of the Road Traffic Act 1988 to give information as to the identity of a driver on a particular occasion to which the complaint relates has been served on the accused by post, and

(b) a statement in writing is produced to the court, purporting to be signed by the accused, that the accused was the driver of that vehicle on that occasion,

that statement shall be sufficient evidence that the accused was the driver of the vehicle on that occasion."

86. In section 14 of that Act (use of records kept by operators of goods vehicles) after the word "proceedings" there shall be inserted the words "for an offence under section 40A of the Road Traffic Act 1988 or".

87.—(1) Section 15 of that Act (use of specimens in proceedings for offences under sections 4 and 5 of the Road Traffic Act 1988) shall be amended as follows.

(2) In subsection (1) for the words "section 4 or 5 of the Road Traffic Act 1988 (motor vehicles: drink and drugs)" there shall be substituted the words "section 3A, 4 or 5 of the Road Traffic Act 1988 (driving offences connected with drink or drugs)", and for the words "sections 4 to 10" there shall be substituted the words "sections 3A to 10".

(3) In subsection (2) after the word "cases" there shall be inserted the words "(including cases where the specimen was not provided in connection with the alleged offence)".

(4) For subsection (3) there shall be substituted—

"(3) That assumption shall not be made if the accused proves—

(a) that he consumed alcohol before he provided the specimen and—

(i) in relation to an offence under section 3A, after the time of the alleged offence, and

(ii) otherwise, after he had ceased to drive, attempt to drive or be in charge of a vehicle on a road or other public place, and

(b) that had he not done so the proportion of alcohol in his breath, blood or urine would not have exceeded the prescribed limit and, if it is alleged that he was unfit to drive through drink, would not have been such as to impair his ability to drive properly."

88.—(1) Section 17 of that Act (provisions as to proceedings for certain offences in connection with the construction and use of vehicles) shall be amended as follows.

(2) In subsection (1) for the words "section 42(1) of the Road Traffic Act 1988 (contravention" there shall be substituted the words "section 40A, 41A, 41B or

42 of the Road Traffic Act 1988 (using vehicle in dangerous condition or contravention".

(3) In subsection (3) after the word "requirements" there shall be inserted the words ", or so that it has ceased to be excessive,".

89. In section 21(3) of that Act (evidence of one witness sufficient in Scotland in relation to certain offences) for the words "or 36" there shall be substituted the words ", 36 or 172."

90.—(1) Section 23 of that Act (alternative verdicts in Scotland) shall be amended as follows.

(2) In subsection (1)—

(a) for the words "motor vehicle" there shall be substituted the words "mechanically propelled vehicle"; and

(b) for the word "reckless" there shall be substituted the word "dangerous".

(3) Subsection (2) shall be omitted.

91.—(1) Section 27 of that Act (production of licence) shall be amended as follows.

(2) In subsection (1), for the words from "endorsement" to "Act" there shall be substituted the words "or discretionary disqualification, and a court proposes to make an order disqualifying him or an order under section 44 of this Act, the court must, unless it has already received them,".

(3) Subsection (2) shall be omitted.

(4) In subsection (3), after the words "as required" there shall be inserted the words "under this section or section 44 of the Powers of Criminal Courts Act 1973, or section 223A or 436A of the Criminal Procedure (Scotland) Act 1975".

92.—(1) Section 30 of that Act (modification of penalty points where fixed penalty also in question) shall be amended as follows.

(2) In subsection (1)(a) for the words "obligatory or discretionary disqualification" there shall be substituted the words "obligatory endorsement".

(3) In subsection (2)—

(a) the words "Subject to section 28(2) of this Act" shall be omitted,

(b) in paragraph (a) for "28(1)" there shall be substituted "28", and

(c) in paragraph (b) at the end there shall be added the words "(except so far as they have already been deducted by virtue of this paragraph)".

(4) Subsection (3) shall be omitted.

93. In section 31(1) of that Act (court may take particulars endorsed on licence into account) for the words "obligatory endorsement" there shall be substituted the words "obligatory or discretionary disqualification".

94. In section 32(1) of that Act (court in Scotland may take extract from licensing records into account) for the words "obligatory endorsement" there shall be substituted the words "obligatory or discretionary disqualification".

95.—(1) Section 35 of that Act (disqualification for repeated offences) shall be amended as follows.

(2) In subsection (1)(a) for the words "involving obligatory or discretionary disqualification" there shall be substituted the words "to which this subsection applies".

(3) After subsection (1) there shall be inserted—

"(1A) Subsection (1) above applies to—

(a) an offence involving discretionary disqualification and obligatory endorsement, and

(b) an offence involving obligatory disqualification in respect of which no order is madc under section 34 of this Act."

(4) In subsection (2) for the words "was imposed" there shall be substituted the words "was for a fixed period of 56 days or more and was imposed".

(5) In subsection (3) for the words "involving obligatory or discretionary disqualification" there shall be substituted the words "to which subsection (1) above applies".

(6) In subsection (5) for the words following "1973" there shall be substituted the words "or section 223A or 436A of the Criminal Procedure (Scotland) Act 1975 (offences committed by using vehicles) or a disqualification imposed in respect of an offence of stealing a motor vehicle, an offence under section 12 or 25 of the Theft Act 1968, an offence under section 178 of the Road Traffic Act 1988, or an attempt to commit such an offence".

(7) After subsection (5) there shall be inserted—

"(5A) The preceding provisions of this section shall apply in relation to a conviction of an offence committed by aiding, abetting, counselling, procuring, or inciting to the commission of, an offence involving obligatory disqualification as if the offence were an offence involving discretionary disqualification."

96. In section 37(3) of that Act (driver disqualified until test is passed entitled to provisional licence) for "36(1)" there shall be substituted "36".

97. After section 41 of that Act there shall be inserted—

41A. "Suspension of disqualification pending determination of applications under section 34B.

(1) Where a person makes an application to a court under section 34B of this Act, the court may suspend the disqualification to which the application relates pending the determination of the application.

(2) Where a court exercises its power under subsection (1) above it must send notice of the suspension to the Secretary of State.

(3) The notice must be sent in such manner and to such address, and must contain such particulars, as the Secretary of State may determine."

98. In section 42 of that Act (removal of disqualification) after subsection (5) there shall be inserted—

"(5A) Subsection (5)(a) above shall apply only where the disqualification was imposed in respect of an offence involving obligatory endorsement; and in any other case the court must send notice of the order made under this section to the Secretary of State.

(5B) A notice under subsection (5A) above must be sent in such manner and to such address, and must contain such particulars, as the Secretary of State may determine."

99.—(1) Section 45 of that Act (effect of endorsement) shall be amended as follows.

(2) In subsection (5)(b), for sub-paragraph (ii) there shall be substituted—

"(ii) an order is made for the disqualification of the offender under section 35 of this Act".

(3) In subsection (6) for the word "reckless" in both places where it occurs there shall be substituted the word "dangerous".

(4) In subsection (7), for paragraph (a) there shall be substituted—

"(a) section 3A, 4(1) or 5(1)(a) of that Act (driving offences connected with drink or drugs), or".

100.—(1) Section 47 of that Act (supplementary provisions as to disqualifications and endorsements) shall be amended as follows.

(2) In subsection (2), for the words from "and, if it" to "disqualified, must" there shall be substituted the words ", and where a court orders the holder of a licence to be disqualified for a period of 56 days or more it must,".

(3) In subsection (3), for the words "any such order" there shall be substituted the words "an order for the endorsement of a licence or the disqualification of a person".

101. For section 48 of that Act (exemption from disqualification and endorsement for offences against construction and use regulations) there shall be substituted—

48. "Exemption from disqualification and endorsement for certain construction and use offences.

(1) Where a person is convicted of an offence under section 40A of the Road Traffic Act 1988 (using vehicle in dangerous condition etc) the court must not—

(a) order him to be disqualified, or

(b) order any particulars or penalty points to be endorsed on the counterpart of any licence held by him,

if he proves that he did not know, and had no reasonable cause to suspect, that the use of the vehicle involved a danger of injury to any person.

(2) Where a person is convicted of an offence under section 41A of the Road Traffic Act 1988 (breach of requirement as to brakes, steering-gear or tyres) the court must not—

(a) order him to be disqualified, or

(b) order any particulars or penalty points to be endorsed on the counterpart of any licence held by him,

if he proves that he did not know, and had no reasonable cause to suspect, that the facts of the case were such that the offence would be committed.

(3) In relation to licences which came into force before 1st June 1990, the references in subsections (1) and (2) above to the counterpart of a licence shall be construed as references to the licence itself."

102. For section 53 of that Act there shall be substituted—

53. "Amount of fixed penalty.

(1) The fixed penalty for an offence is—

(a) such amount as the Secretary of State may by order prescribe, or

(b) one half of the maximum amount of the fine to which a person committing that offence would be liable on summary conviction,

whichever is the less.

(2) Any order made under subsection (1)(a) may make different provision for different cases or classes of case or in respect of different areas."

103.—(1) Section 54 of that Act (power to give fixed penalty notices on the spot or at a police station exercisable only if offender would not if convicted be liable to disqualification under section 35) shall be amended as follows.

(2) In subsection (1), after the word "where", there shall be inserted the words "in England and Wales".

(3) After subsection (9) there shall be added—

"(10) In determining for the purposes of subsections (3)(b) and (5)(a) above whether a person convicted of an offence would be liable to disqualification under section 35, it shall be assumed, in the case of an offence in relation to which a range of numbers is shown in the last column of Part I of Schedule 2 to this Act, that the number of penalty points to be attributed to the offence would be the lowest in the range."

104. In section 61 of that Act (fixed penalty notice mistakenly given) after subsection (5) there shall be added—

"(6) In determining for the purposes of subsection (1) above whether a person convicted of an offence would be liable to disqualification under section 35, it shall be assumed, in the case of an offence in relation to which a range of numbers is shown in the last column of Part I of Schedule 2 to this Act, that the number of penalty points to be attributed to the offence would be the lowest in the range."

105. In section 69(4) of that Act (references to fixed penalty clerk) after the words "of this Act" there shall be inserted the words "except in sections 75 to 77)".

106. At the end of section 86(1) of that Act (functions of traffic wardens) there shall be added the words "unless that offence was committed whilst the vehicle concerned was stationary."

107.—(1) Section 89 of that Act (interpretation), shall be amended as follows.

(2) After the definition of "authorised person" there shall be inserted—

"'chief constable' means, in Scotland in relation to any conditional offer, the chief constable for the area in which the conditional offer has been issued."

(3) In the definition of "chief officer of police" for the words "or notice to owner" there shall be substituted the words ", notice to owner or conditional offer".

108. In section 90 of that Act (index to Part III)—

(a) in the entry relating to the expression "Conditional offer", for the words "Section 75(4)" there shall be substituted the words "Section 75(3)"; and

(b) at the end of the entry relating to the expression "Fixed penalty clerk" there shall be added the words "and 75(4)".

109. In section 92 of that Act (persons in public service of Crown) after "16" there shall be inserted"20".

110. In section 93(2) of that Act (persons subject to service discipline for "4" there shall be substituted "3A".

111.—(1) In subsection (1) of section 98 of that Act (interpretation), in paragraph (b) of the definition of "road", for the words "has the same meaning as in the Roads (Scotland) Act 1984" there shall be substituted the words "means any road within the meaning of the Roads (Scotland) Act 1984 and any other way to which the public has access, and includes bridges over which a road passes,".

(2) At the end of subsection (2) of that section there shall be added the word "Act".

112.—(1) Schedule 3 to the Road Traffic Offenders Act 1988 (fixed penalty offences) shall be amended as follows.

(2) After the entry relating to section 36 of the Road Traffic Act 1988 there shall be inserted—

"RTA section 40A	Using vehicle in dangerous condition etc.
RTA section 41A	Breach of requirement as to brakes, steering-gear or tyres.
RTA section 41B	Breach of requirement as to weight: goods and passenger vehicles."

(3) In the entry relating to section 42 of the Road Traffic Act 1988, for the words in column 2 there shall be substituted the words "Breach of other construction and use requirements".

(4) In the entry relating to section 87(1) of the Road Traffic Act 1988, in column 2 for the word "without" there shall be substituted the words "otherwise than in accordance with".

113. In Schedule 5 to the Road Traffic Offenders Act 1988 (conditional offer of fixed penalty in relation to certain offences in Scotland), in the entry relating to section 87(2) of the Road Traffic Act 1988, in column (2) for the word "without" there shall be substituted the words "otherwise than in accordance with".

The Road Traffic (Consequential Provisions) Act 1988 (c. 54)

114. In section 8(3) of the Road Traffic (Consequential Provisions) Act 1988 (commencement) for the words from the beginning to the end of paragraph (c) there shall be substituted the words "Paragraphs 15 to 20 of Schedule 2 to this Act".

Section 52

SCHEDULE 5
THE TRAFFIC DIRECTOR FOR LONDON

Status

1. The Traffic Director for London shall be a corporation sole.

2. The Director shall not be regarded as the servant or agent of the Crown or as enjoying any status, immunity or privilege of the Crown; and the Director's property shall not be regarded as property of, or held on behalf of, the Crown.

Tenure of office

3.—(1) Subject to the following provisions of this paragraph, the Director shall hold and vacate office in accordance with the terms of his appointment.

(2) The Director shall be appointed for a term not exceeding five years.

(3) At the end of a term of appointment the Director shall be eligible for re-appointment.

(4) The Director may at any time resign his office by notice in writing addressed to the Secretary of State.

(5) The Secretary of State may remove the Director from office—

(a) if a bankruptcy order has been made against him, or his estate has been sequestrated or he has made a composition or arrangement with, or granted a trust deed for, his creditors; or

(b) if satisfied that he is otherwise unable or unfit to discharge his functions.

(6) The Director's terms of appointment may provide for his removal from office (without assigning cause) on notice from the Secretary of State of such length as may be specified in those terms, subject, if those terms so provide, to compensation of such amount as the Secretary of State may, with the approval of the Treasury, determine.

Remuneration etc

4.—(1) There shall be paid to the Director such remuneration and such travelling and other allowances as the Secretary of State may determine.

(2) In the case of any such holder of the office of Director as may be determined by the Secretary of State, there shall be paid such pension, allowances or gratuities to or in respect of him, or such payments towards the provision of a pension, allowances or gratuities to or in respect of him, as may be so determined.

(3) If the Secretary of State determines that there are special circumstances which make it right that a person ceasing to hold office as Director should receive compensation, there may be paid to him a sum by way of compensation of such amount as the Secretary of State may determine.

(4) Sub-paragraph (3) above does not apply in the case of a person who receives compensation by virtue of paragraph 3(6) above.

(5) The approval of the Treasury shall be required for the making of a determination under this paragraph.

Staff

5.—(1) The Director shall act only with the approval of the Secretary of State, given with the approval of the Treasury, in determining—

(a) the number of persons to be employed by him;

(b) the remuneration, allowances and gratuities to be paid to or in respect of such persons; and

(c) any other terms and conditions of their service.

(2) Anything authorised or required by or under any enactment to be done by the Director may be done by any person employed by him who has been authorised by the Director, whether generally or specially, for that purpose.

(3) Employment by the Director shall be included among the kinds of employment to which a scheme under section 1 of the Superannuation Act 1972 may apply; and, accordingly, in Schedule 1 to that Act (in which those kinds of employment are listed) at the end of the list of "Other Bodies" there shall be inserted—

"Employment by the Traffic Director for London."

(4) The Director shall pay to the Treasury, at such times as the Treasury may direct, such sums as the Treasury may determine in respect of the increase in the sums payable out of money provided by Parliament under that Act attributable to sub-paragraph (3) above.

(5) Where an employee of the Director who is (by reference to that employment) a participant in a scheme under section 1 of that Act, becomes a holder of the office of Director, the Treasury may determine that his term of office shall be treated for the purposes of the scheme as employment by the Director (whether or not any benefits are payable to or in respect of him by virtue of paragraph 4(2) above).

Financial provisions

6. The remuneration of the Director and any other payments made under paragraphs 3(6) or 4 above to or in respect of him shall be paid out of grants made by the Secretary of State under section 80(2) of this Act.

Accounts

7.—(1) The Director shall keep accounts and shall prepare a statement of accounts in respect of each financial year.

(2) The accounts shall be kept, and the statement shall be prepared, in such form as the Secretary of State may, with the approval of the Treasury, direct.

(3) The accounts shall be audited by persons appointed in respect of each financial year by the Secretary of State.

(4) No person shall be qualified to be appointed as auditor under this paragraph unless he is—

(a) a member of a body of accountants established in the United Kingdom and for the time being recognised for the purposes of Part II of the Companies Act 1989; or

(b) a member of the Chartered Institute of Public Finance and Accountancy;

but a firm may be appointed as auditor under this paragraph if each of its members is qualified to be so appointed.

(5) In this paragraph, and in paragraph 8 below, "financial year" means—

(a) the period beginning with the day on which the first person to hold the office of Director takes office and ending with the following 31st March; and

(b) each subsequent period of twelve months ending with 31st March.

Annual report etc

8.—(1) As soon as possible after the end of each financial year, the Director shall submit to the Secretary of State an annual report on the discharge in that year of his functions.

(2) Each report shall contain a copy of the statement of accounts prepared and audited under paragraph 7 above in respect of that financial year.

(3) The Secretary of State shall lay a copy of the Director's annual report before each House of Parliament.

(4) The Director shall provide the Secretary of State with such information relating to his property and the discharge and proposed discharge of his functions as the Secretary of State may require; and for that purpose shall—

(a) permit any person authorised in that behalf by the Secretary of State to make copies of any accounts or other documents; and

(b) give such explanation as may be required of any such accounts or documents.

Evidence

9. A document purporting to be duly executed under the seal of the Director or to be signed on the Director's behalf shall be received in evidence and, unless the contrary is proved, be deemed to be so executed or signed.

Public records

10. In Schedule 1 to the Public Records Act 1958, in Part II of the Table in paragraph 3 the following entry shall be inserted at the appropriate place—
"Traffic Director for London".

The Parliamentary Commissioner

11. In the Parliamentary Commissioner Act 1967, in Schedule 2 (departments and authorities subject to investigation), the following entry shall be inserted at the appropriate place—
"Traffic Director for London."

Parliamentary disqualification

12.—(1) In the House of Commons Disqualification Act 1975, in Part III of Schedule 1 (other disqualifying offices), the following entry shall be inserted at the appropriate place—
"Traffic Director for London."

(2) The same entry shall be inserted at the appropriate place in Part III of Schedule 1 to the Northern Ireland Assembly Disqualification Act 1975.

Section 66(7).

SCHEDULE 6
PARKING PENALTIES

The notice to owner

1.—(1) Where—

(a) a penalty charge notice has been issued with respect to a vehicle under section 66 of this Act; and

(b) the period of 28 days for payment of the penalty charge has expired without that charge being paid,

the London authority concerned may serve a notice ("a notice to owner") on the person who appears to them to have been the owner of the vehicle when the alleged contravention occurred.

(2) A notice to owner must state—

(a) the amount of the penalty charge payable;

(b) the grounds on which the parking attendant who issued the penalty charge notice believed that a penalty charge was payable with respect to the vehicle;

(c) that the penalty charge must be paid before the end of the period of 28 days beginning with the date on which the notice to owner is served;

(d) that failure to pay the penalty charge may lead to an increased charge being payable;

(e) the amount of that increased charge;

(f) that the person on whom the notice is served ("the recipient") may be entitled to make representations under paragraph 2 below; and

(g) the effect of paragraph 5 below.

(3) The Secretary of State may prescribe additional matters which must be dealt with in any notice to owner.

Representations against notice to owner

2.—(1) Where it appears to the recipient that one or other of the grounds mentioned in sub-paragraph (4) below are satisfied, he may make representations to that effect to the London authority who served the notice on him.

(2) Any representations under this paragraph must be made in such form as may be specified by the London authorities, acting through the Joint Committee.

(3) The authority may disregard any such representations which are received by them after the end of the period of 28 days beginning with the date on which the notice to owner was served.

(4) The grounds are—

(a) that the recipient—

(i) never was the owner of the vehicle in question;

(ii) has ceased to be its owner before the date on which the alleged contravention occurred; or

(iii) became its owner after that date;

(b) that the alleged contravention did not occur;

(c) that the vehicle had been permitted to remain at rest in the parking place by a person who was in control of the vehicle without the consent of the owner;

(d) that the relevant designation order is invalid;

(e) that the recipient is a vehicle-hire firm and—

(i) the vehicle in question was at the material time hired from that firm under a vehicle hiring agreement; and

(ii) the person hiring it had signed a statement of liability acknowledging his liability in respect of any penalty charge notified fixed to the vehicle during the currency of the hiring agreement;

(f) that the penalty charge exceeded the amount applicable in the circumstances of the case.

(5) Where the ground mentioned in sub-paragraph (4)(a)(ii) above is relied on in any representations made under this paragraph, those representations must include a statement of the name and address of the person to whom the vehicle was disposed of by the person making the representations (if that information is in his possession).

(6) Where the ground mentioned in sub-paragraph (4)(a)(iii) above is relied on in any representations made under this paragraph, those representations must include a statement of the name and address of the person from whom the vehicle was acquired by the person making the representations (if that information is in his possession).

(7) It shall be the duty of an authority to whom representations are duly made under this paragraph—

(a) to consider them and any supporting evidence which the person making them provides; and

(b) to serve on that person notice of their decision as to whether they accept that the ground in question has been established.

Cancellation of notice to owner

3.—(1) Where representations are made under paragraph 2 above and the London authority concerned accept that the ground in question has been established they shall—

(a) cancel the notice to owner; and

(b) state in the notice served under paragraph 2(7) above that the notice to owner has been cancelled.

(2) The cancellation of a notice to owner under this paragraph shall not be taken to prevent the London authority concerned serving a fresh notice to owner on another person.

(3) Where the ground that is accepted is that mentioned in paragraph 2(4)(e) above, the person hiring the vehicle shall be deemed to be its owner for the purposes of this Schedule.

Rejection of representations against notice to owner

4. Where any representations are made under paragraph 2 above but the London authority concerned do not accept that a ground has been established, the notice served under paragraph 2(7) above ("the notice of rejection") must—

(a) state that a charge certificate may be served under paragraph 6 below unless before the end of the period of 28 days beginning with the date of service of the notice of rejection—

(i) the penalty charge is paid; or

(ii) the person on whom the notice is served appeals to a parking adjudicator against the penalty charge;

(b) indicate the nature of a parking adjudicator's power to award costs against any person appealing to him; and

(c) describe in general terms the form and manner in which an appeal to a parking adjudicator must be made,

and may contain such other information as the authority consider appropriate.

Adjudication by parking adjudicator

5.—(1) Where an authority serve notice under sub-paragraph (7) of paragraph 2 above, that they do not accept that a ground on which representations were made under that paragraph has been established, the person making those representations may, before—

(a) the end of the period of 28 days beginning with the date of service of that notice; or

(b) such longer period as a parking adjudicator may allow,

appeal to a parking adjudicator against the authority's decision.

(2) On an appeal under this paragraph, the parking adjudicator shall consider the representations in question and any additional representations which are made by the appellant on any of the grounds mentioned in paragraph 2(4) above and may give the London authority concerned such directions as he considers appropriate.

(3) It shall be the duty of any authority to whom a direction is given under sub-paragraph (2) above to comply with it forthwith.

Charge certificates

6.—(1) Where a notice to owner is served on any person and the penalty charge to which it relates is not paid before the end of the relevant period, the authority serving the notice may serve on that person a statement (a "charge certificate") to the effect that the penalty charge in question is increased by 50 per cent.

(2) The relevant period, in relation to a notice to owner, is the period of 28 days beginning—

(a) where no representations are made under paragraph 2 above, with the date on which the notice to owner is served;

(b) where—

(i) such representations are made;

(ii) a notice of rejection is served by the authority concerned; and

(iii) no appeal against the notice of rejection is made,

with the date on which the notice of rejection is served; or

(c) where there has been an unsuccessful appeal against a notice of rejection, with the date on which notice of the adjudicator's decision is served on the appellant.

(3) Where an appeal against a notice of rejection is made but is withdrawn before the adjudicator gives notice of his decision, the relevant period in relation to a notice to owner is the period of 14 days beginning with the date on which the appeal is withdrawn.

Enforcement of charge certificate

7. Where a charge certificate has been served on any person and the increased penalty charge provided for in the certificate is not paid before the end of the period of 14 days beginning with the date on which the certificate is served, the authority concerned may, if a county court so orders, recover the increased charge as if it were payable under a county court order.

Invalid notices

8.—(1) This paragraph applies where—

(a) a county court makes an order under paragraph 7 above;

(b) the person against whom it is made makes a statutory declaration complying with sub-paragraph (2) below; and

(c) that declaration is, before the end of the period of 21 days beginning with the date on which notice of the county court's order is served on him, served on the county court which made the order.

(2) The statutory declaration must state that the person making it—

(a) did not receive the notice to owner in question;

(b) made representations to the London authority concerned under paragraph 2 above but did not receive a rejection notice from that authority; or

(c) appealed to a parking adjudicator under paragraph 5 above against the rejection by that authority of representations made by him under paragraph 2 above but had no response to the appeal.

(3) Sub-paragraph (4) below applies where it appears to a district judge, on the application of a person on whom a charge certificate has been served, that it would be unreasonable in the circumstances of his case to insist on him serving his statutory declaration within the period of 21 days allowed for by sub-paragraph (1) above.

(4) Where this sub-paragraph applies, the district judge may allow such longer period for service of the statutory declaration as he considers appropriate.

(5) Where a statutory declaration is served under sub-paragraph (1)(c) above—

(a) the order of the court shall be deemed to have been revoked;

(b) the charge certificate shall be deemed to have been cancelled;

(c) in the case of a declaration under sub-paragraph (2)(a) above, the notice to owner to which the charge certificate relates shall be deemed to have been cancelled; and

(d) the district judge shall serve written notice of the effect of service of the declaration on the person making it and on the London authority concerned.

(6) Service of a declaration under sub-paragraph (2)(a) above shall not prevent the London authority serving a fresh notice to owner.

(7) Where a declaration has been served under sub-paragraph (2)(b) or (c) above, the London authority shall refer the case to the parking adjudicator who may give such direction as he considers appropriate.

Offence of giving false information

9.—(1) A person who, in response to a notice to owner served under this Schedule, makes any representation under paragraph 2 or 5(2) above which is false in a material particular and does so recklessly or knowing it to be false in that particular is guilty of an offence.

(2) Any person guilty of such an offence shall be liable on summary conviction to a fine not exceeding level 5 on the standard scale.

Service by post

10. Any charge certificate, or notice under this Schedule—

(a) may be served by post; and

(b) where the person on whom it is to be served is a body corporate, is duly served if it is sent by post to the secretary or clerk of that body.

Section 81.

SCHEDULE 7
MINOR AND CONSEQUENTIAL AMENDMENTS IN RELATION TO LONDON

The Tribunals and Inquiries Act 1971 (c. 62)

1. In paragraph 30 of the Tribunals and Inquiries Act 1971, in Part I of Schedule 1 (tribunals under direct supervision of the Council on Tribunals) after "30" there shall be inserted "(a)", and at the end of that paragraph there shall be inserted the words "and

(b) a parking adjudicator appointed under section 73(3)(a) of the Road Traffic Act 1991."

The Greater London Council (General Powers) Act 1974 (c. xxiv)

2. In section 15 of the Greater London Council (General Powers) Act 1974 (parking on footways etc.) in subsection (12)(b) for the words "under section 84" there shall be substituted the words "made by virtue of section 84(1)(a)".

The Road Traffic Regulation Act 1984 (c. 27)

3. In section 7 of the Road Traffic Regulation Act 1984 (provisions supplementary to section 6), in subsection (6) for the words "Secretary of State for the Home Department" there shall be substituted the words "the Commissioner of Police for any police area in which is situated any road or part of a road to which the order is to relate".

4. In that Act, after section 13 there shall be inserted—

"Temporary suspension

13A. Temporary suspension of provisions under s. 6 or 9 orders.

(1) The Commissioner of Police of the Metropolis or the Commissioner of Police for the City of London may temporarily suspend the operation of any provision of an order made under section 6 or 9 of this Act so far as that provision relates to any road or part of a road in Greater London which is within his area, in order to prevent or mitigate congestion or obstruction of traffic, or danger to or from traffic in consequence of extraordinary circumstances.

(2) Subject to subsection (3) below, the period of suspension under subsection (1) above shall not continue for more than 7 days.

(3) If the Secretary of State gives his consent to the period of suspension being continued for more than 7 days, the suspension shall continue until the end of such period as may be specified by the Secretary of State in giving his consent."

5.—(1) Section 55 of that Act (financial provisions relating to designation orders) shall be amended as follows.

(2) In subsection (1), for the words from "designated" to the end there shall be substituted the words "for which they are the local authority and which are—

(a) in the case of the council of a London borough and the Common Council of the City of London, parking places on the highway; and

(b) in the case of any other authority, designated parking places."

(3) After subsection (3) there shall be inserted—

"(3A) The council of each London borough and the Common Council of the City of London shall, after each financial year, report to the Secretary of State on any action taken by them, pursuant to subsection (2) or (3) above, in respect of any deficit or surplus in their account for the year.

(3B) The report under subsection (3A) above shall be made as soon after the end of the financial year to which it relates as is reasonably possible."

(4) In subsection (4)(c), the words from "to the council" to "City of London" shall be omitted.

6.—(1) Section 105 of that Act (exemptions from provisions relating to immobilisation of vehicles) shall be amended as follows.

(2) In subsection (2) after the words "of any vehicle" there shall be inserted the words "found otherwise than in Greater London".

(3) After subsection (2) there shall be inserted—

"(2A) The exemption under subsection (1)(b) above shall not apply in the case of any vehicle found in Greater London if the meter bay in which it was found was not authorised for use as such at the time when it was left there."

(4) In subsection (3) for the words "subsection (2)(a)" there shall be substituted the words "subsections (2)(a) and (2A)".

7. In section 122 of that Act (exercise of functions by local authorities) there shall be added at the end—

"(3) The duty imposed by subsection (1) above is subject to the provisions of Part II of the Road Traffic Act 1991."

The Local Government Act 1985 (c. 51)

8.—(1) For paragraph 5 of Schedule 5 to the Local Government Act 1985 (designation of routes in London) there shall be substituted—

"5.—(1) For the purpose of facilitating the movement of traffic in Greater London, the Secretary of State may by order designate a road in that area.

(2) Before doing so, he shall consult—

(a) the council of the London borough in which the road is;

(b) the council of any other London borough or of any county where there is a road which he considers is likely to be affected by the designation; and

(c) such other persons (if any) as he considers it appropriate to consult.

(3) No council of a London borough shall exercise any power under the Highways Act 1980 or the Road Traffic Regulation Act 1984 in a way which will affect, or be likely to affect, a designated road unless the requirements of sub-paragraph (4) below have been satisfied.

(4) The requirements are that—

(a) the council concerned has given notice to the Director, in such manner as he may require, of its proposal to exercise the power in the way in question; and

(b) either—

(i) the Director has approved the proposal; or

(ii) the period of one month beginning with the date on which he received notice of the proposal has expired without his having objected to it.

(5) The Secretary of State may by an instrument in writing exclude any power from the application of this paragraph to the extent specified in the instrument.

(6) Any such instrument may, in particular, exclude a power as respects—

(a) all or any of the London boroughs;

(b) all or any of the designated roads; or

(c) the exercise of the power in such manner or circumstances as may be specified in the instrument.

(7) This paragraph does not apply to the exercise of a power under section 14 or section 32 to 38 of the 1984 Act in relation to a road which is not a designated road.

(8) If the council of a London borough exercises any power in contravention of this paragraph, the Director may take such steps as he considers appropriate to reverse or modify the effect of the exercise of that power.

(9) Any reasonable expenses incurred by the Director in taking any steps under sub-paragraph (8) shall be recoverable by him from the council as a civil debt.

(10) In this paragraph—

"designated road" means a road designated under this paragraph; and

"Director" means the Traffic Director for London."

9.—(1) Paragraph 6 of that Schedule (guidance as to exercise of traffic powers) shall be amended as follows.

(2) In sub-paragraph (5) for the words "summary as a civil debt" there shall be substituted the words "as a debt due to the Crown".

(3) After sub-paragraph (6), there shall be inserted—

"(7) Sub-paragraphs (3) to (6) above shall not apply in relation to the exercise of any power, by the council of a London borough, in complying with the duty imposed on them by section 57(1) of the Road Traffic Act 1991 (implementation of local plans)."

10. In paragraph 10(6) of that Schedule (recovery of sums expended by the Secretary of State in connection with traffic control systems) for the words "summarily as a civil debt" there shall be substituted the words "as a debt due to the Crown".

11. In paragraph 11 of that Schedule (recovery of sums expended by the Secretary of State to obtain information) for the words "summarily as a civil debt" there shall be substituted the words "as a debt due to the Crown".

The New Roads and Street Works Act 1991 (c. 22)

12. In section 64 of the New Roads and Street Works Act 1991 (traffic-sensitive streets), after subsection (3) there shall be added—

"(4) Where any council of a London borough or the Common Council of the City of London are asked by the Traffic Director for London to designate a street as a traffic-sensitive street and they decline to do so, the Director may appeal to the Secretary of State who may direct that the street be designated."

Section 83.

SCHEDULE 8
REPEALS

Chapter	Short title	Extent of repeal
1970 c. 44	The Chronically Sick and Disabled Persons Act 1970.	In section 21(4) the words "and any badge" onwards. In section 21(5) the words "and in the case" onwards.
1972 c. 27.	The Road Traffic (Foreign Vehicles) Act 1972.	In Schedule 1— the entry relating to section 8(1) of the Public Passenger Vehicles Act 1981; in the entry relating to section 68 of the Road Traffic Act 1988, the word "goods".
1972 c. 71.	The Criminal Justice Act 1972.	Section 24(2).
1973 c. 62.	The Powers of Criminal Courts Act 1973.	In section 44(3), paragraphs (a) and (b) and the word "and" immediately preceding them.
1975 c. 46.	The International Road Haulage Permits Act 1975.	In section 1(9), the words "section 56(1) of the Road Traffic Act 1972 or".
1981 c. 14.	The Public Passenger Vehicles Act 1981.	Section 7. Section 8(1) to (2). In section 8(3), the words "for the purposes of this Act". Section 9. In section 9A(1), the words "with the omission of subsection (1)(b)". Section 9A(2). Section 20(6). Section 51(2). In section 53(1), the words "certifying officers, public service vehicle examiners" and the words "public service" in the second place where they appear. Section 65(1)(f). In section 66A(1), paragraph (b) and the word "or" immediately preceding it. In section 68(4), the reference to section 9(9)(b). In section 82(1), the definition of "certifying officer". In Schedule 7, paragraph 17(a).

Chapter	Short title	Extent of repeal
1982 c. 49.	The Transport Act 1982.	In section 9, the paragraph beginning "Any functions under section 9". Section 10(5). In section 10(8), the words from "Without prejudice" to "their functions". Section 19. Section 21(2) and (3). Section 23(4). In Schedule 5, paragraph 21.
1984 c. 27.	The Road Traffic Regulation Act 1984.	In section 17(2), the word "or" at the end of paragraph (b). Section 35(9). In section 51(5), the words "being not less than 2 years". In section 55(4)(c), the words from "to the Council" to "City of London". In section 85(1), the words "the prescribed". In section 85(2)(a), the words "the prescribed". In section 99(2), paragraph (c) and the word "and" immediately preceding it. In section 102(2), the word "and" at the end of paragraph (b). In section 102(8), the words following paragraph (b) in the definition of "appropriate authority", and the word "and" at the end of the definition of "person responsible". Section 104(10). In section 105(3)(b), the words "under section 49(4) of this Act." In section 106— subsections (2) to (4), (6) and (10); in subsection (5), the words "After the end of the experimental period"; in subsection (9), the words "except in the case of an order to which subsection (6) above applies". In section 117(3), the definition of "disabled person's badge". Section 141.

Chapter	Short title	Extent of repeal
		In Schedule 13, in paragraph 40, the words "and for" onwards.
1985 c. 67.	The Transport Act 1985.	In Part II of Schedule 2, paragraph 4(3) and (11)(b). In Schedule 7, paragraph 21(2) and (3).
1988 c. 52.	The Road Traffic Act 1988.	Section 15(10). Section 19A. In section 29, the words "In this section" to the end. Section 30(3). Section 41(3)(b) and (c). Section 48(6). Section 50(2) and (3). In section 51(1)(b), the word "goods". In section 61(2)(a), the words from "goods" to "service". Section 61(5). Section 67(4)(a). Section 73(2). Section 75(3)(a)(iii). In section 75(6), paragraph (c) and the word "or" immediately preceding it. Section 75(8). Section 79(2)(a). In section 86, in the table, the entry for "Goods vehicle examiner". Section 97(7). Section 98(5). In section 105(2)(ee), the words "for any purpose of this Part of this Act". In section 105(2)(f), the words "for the purposes of this Part of this Act". Section 151(9)(b). In section 164(6), the words "to a constable". In section 165(4), the words "to a constable". In section 173(2), the word "and" at the end of paragraph (k). In section 183(3), paragraph (b) and the word "and" immediately preceding it. In section 192(2), the word ""road"". Section 193. Schedule 4.

Chapter	Short title	Extent of repeal
1988 c. 53.	The Road Traffic Offenders Act 1988.	In section 17(3), the word "goods" in each place where it occurs. Section 23(2). Section 27(2). In section 30(2), the words "Subject to section 28(2) of this Act,". Section 30(3). Section 54(8). Section 59(6). Section 60. In Schedule 1, in the Table— in the entry relating to section 71 of the Road Traffic Act 1988, in column 2 the word "goods" in each place where it occurs; the entries relating to section 97 and 98 of that Act. In Part I of Schedule 2— in the entries relating to sections 68 and 71 of the Road Traffic Act 1988, in column 2 the word "goods" in each place where it occurs; the entries relating to sections 97 and 98 of that Act; in the entry relating to section 165 of that Act, in column 2 the word "constable"; the entries in columns 6 and 7 relating to section 178 of that Act; in the entry relating to section 26 of the Road Traffic Offenders Act, in column 2 the words "on committal for sentence etc." In Part II of Schedule 2, the entries in columns 3 and 4 relating to stealing or attempting to steal a motor vehicle or to section 12 or 25 of the Theft Act 1968. In Schedule 3, the entry relating to section 97 of the Road Traffic Act 1988.

Chapter	Short title	Extent of repeal

1988 c. 54.	The Road Traffic (Consequential Provisions) Act 1988.	Section 6.
		In Part I of Schedule 2—
		paragraph 1;
		in paragraph 3(1) the entry beginning "for "section 56(2)(a)"".
		paragraph 4(2);
		paragraph 8;
		paragraph 9;
		paragraph 10(b);
		paragraph 13(b)(ii);
		paragraph 15(b) and the word "and" immediately preceding it.
		Parts II, III and IV of Schedule 2.
		In Schedule 3—
		paragraph 6(3) and (5);
		paragraph 8(1);
		paragraph 8(2)(d) and the word "and" immediately preceding it;
		paragraph 9(1)(c) and the word "and" immediately preceding it;
		paragraph 9(3)(b);
		paragraph 11(b) and (c);
		paragraph 37(1) and (2).
		Schedule 5.

1989 c. 22.	The Road Traffic (Driver Licensing and Information Systems) Act 1989.	In Schedule 3, paragraph 21.
1991 c. 40.	The Road Traffic Act 1991.	In Schedule 4, paragraph 79.

Index